Analytical Tools for Atmospheric Systems

Edited by **Mary D'souza**

SYRAWOOD
PUBLISHING HOUSE

New York

Published by Syrawood Publishing House,
750 Third Avenue, 9ᵗʰ Floor,
New York, NY 10017, USA
www.syrawoodpublishinghouse.com

Analytical Tools for Atmospheric Systems
Edited by Mary D'souza

International Standard Book Number: 978-1-68286-000-7 (Hardback)

Printed in the United States of America.

Contents

Preface

The study of atmospheric systems focuses on the study of all the elements constituting earth's environment. It is an important field of science that has undergone rapid development over the past few decades. This book covers in detail some existent theories and innovative concepts such as application of remote sensing in analysis of different gases and aerosols, measurement of wind, precipitation, temperature, etc. Comprising of state-of-the-art inputs by acclaimed experts of this field, this book targets students and professionals involved in various branches of atmospheric sciences.

Various studies have approached the subject by analyzing it with a single perspective, but the present book provides diverse methodologies and techniques to address this field. This book contains theories and applications needed for understanding the subject from different perspectives. The aim is to keep the readers informed about the progress in the field; therefore, the contributions were carefully examined to compile novel researches by specialists from across the globe.

Indeed, the job of the editor is the most crucial and challenging in compiling all chapters into a single book. In the end, I would extend my sincere thanks to the chapter authors for their profound work. I am also thankful for the support provided by my family and colleagues during the compilation of this book.

Editor

Fiber optic distributed temperature sensing for the determination of air temperature

S. A. P. de Jong, J. D. Slingerland, and N. C. van de Giesen

Faculty of Civil Engineering and Geosciences, Delft University of Technology, Delft, the Netherlands

Correspondence to: N. C. van de Giesen (n.c.vandegiesen@tudelft.nl)

Abstract. This paper describes a method to correct for the effect of solar radiation in atmospheric distributed temperature sensing (DTS) applications. By using two cables with different diameters, one can determine what temperature a zero diameter cable would have. Such a virtual cable would not be affected by solar heating and would take on the temperature of the surrounding air. With two unshielded cable pairs, one black pair and one white pair, good results were obtained given the general consensus that shielding is needed to avoid radiation errors (WMO, 2010). The correlations between standard air temperature measurements and air temperatures derived from both cables of colors had a high correlation coefficient ($r^2 = 0.99$) and a RMSE of 0.38 °C, compared to a RMSE of 2.40 °C for a 3.0 mm uncorrected black cable. A thin white cable measured temperatures that were close to air temperature measured with a nearby shielded thermometer (RMSE of 0.61 °C). The temperatures were measured along horizontal cables with an eye to temperature measurements in urban areas, but the same method can be applied to any atmospheric DTS measurements, and for profile measurements along towers or with balloons and quadcopters.

1 Introduction

Distributed temperature sensing (DTS) is a technique that allows for measurement of temperature along optical fibers. Laser pulses are shot into the fiber, and backscatter from within the fiber is analyzed. The time of flight then gives the position along the fiber from where the backscatter originated. Analysis of the Raman spectrum of the backscatter allows for the calculation of the temperature at the place where the backscatter originated. Depending on the type of DTS machine used, temperatures can be measured continuously at sub-meter intervals along cables of more than 5 km, with laboratory accuracies up to 0.01 °C and typical field accuracies of 0.08 °C (van de Giesen et al., 2012). A good introduction to DTS principles and environmental applications can be found in Selker et al. (2006) and Tyler et al. (2009).

Over the past decade, DTS has found many environmental applications. Applications vary from temperature profiling of the subsurface: borehole observations (Freifeld et al., 2008), soils (Ciocca et al., 2012; Jansen et al., 2011; Sayde et al., 2010; Steele-Dunne et al., 2010); water: estuaries (Henderson et al., 2009), surface/groundwater (Lowry et al., 2007; Mamer and Lowry, 2013), solar ponds (Suárez et al., 2011), streams (Selker et al., 2006; Vogt et al., 2010; Westhoff et al., 2007, 2011) and lakes (Vercauteren et al., 2011; van Emmerik et al., 2013); rocks (Read et al., 2013), ice caves (Curtis and Kyle, 2011), forests (Krause et al., 2013) and infrastructure: dam surveillance (Dornstadter, 1998), sewers (Hoes et al., 2009), electric transmission cables (Yilmaz and Karlik, 2006) and gas pipelines (Tanimola and Hill, 2009).

There are only a few experiments where DTS is used to measure atmospheric temperature (Keller et al., 2011; Petrides et al., 2011; Thomas et al., 2012), since solar heating can have a significant effect. Keller et al. (2011) experimented during nighttime to exclude the effect of short-wave radiation. Petrides et al. (2011) estimated effective shade and concluded that solar radiation is the driving factor in temperature differences. Thomas et al. (2012) observed differences in temperature measurements with black and white cables and suggested that it can be used for setting up an energy balance.

This paper will describe a method to correct for the effect of solar radiation in atmospheric DTS measurements with

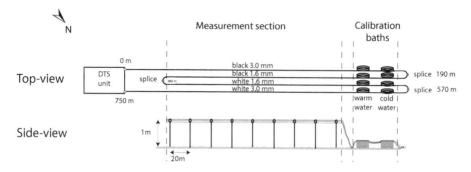

Figure 1. Schematic overview of the experimental setup. The cable exists of four sections (black with diameter 3.0 and 1.6 mm and white with diameter 3.0 and 1.6 mm). In the measurement section the cables are held in open air above a grass field.

the use of fiber optic cables with different diameters. The original objective of the experiment was to test whether air temperature differences in urban landscapes can be measured with reasonable accuracies (1 °C).

2 Materials and methods

2.1 Theory

Solar radiation causes objects to be warmer than the surrounding air. For this reason, thermometers are traditionally shielded by a screen. For accurate measurements, forced ventilation is necessary, which was not provided in our case (WMO, 2010). The temperature difference between a heated cylinder (or sphere) and the air that moves around it scales with the square root of the diameter (White, 1988). If the diameter of a cylinder were zero, the heat generated by solar radiation would also be zero. Such a zero diameter cylinder would take on the temperature of the surrounding air. One can create a virtual cylinder with zero diameter by extrapolating the temperatures of two cylinders with different diameters. The fiber optic cables used by DTS can be considered to be very long cylinders. When the temperatures of two cables (T_1 and T_2) with different diameters (d_1 and d_2) are measured, then the air temperature can be determined by extrapolating to a zero diameter, which results in Eq. (1):

$$T_{\text{air}} = T_2 - \frac{T_1 - T_2}{\sqrt{\frac{d_1}{d_2}} - 1}. \tag{1}$$

This idea (G. Campbell, personal communication, 2010) is based on the assumptions of having an infinitely long cylinder, instant redistribution of heat within a cross section of the cylinder, and forced convection on the outside of the cylinder. Forced convection dominates when the buoyancy force parameter, also known as the Archimedes number, Ar, is much smaller than one. For a cylinder of diameter d (m) we have

$$\text{Ar} = \frac{\text{Gr}}{\text{Re}^2} = g \cdot \frac{T_s - T_{\text{air}}}{T_{\text{air}}} \cdot \frac{d}{v^2}, \tag{2}$$

where Gr is the Grashof number (dimensionless), Re the Reynolds number (dimensionless), g the acceleration due to gravity (9.8 m s^{-2}), T_{air} the air temperature (K), T_s the surface temperature of the cable (K) and v the wind speed (m s^{-1}).

2.2 Experimental setup

The measurements were taken from 27 April 2011 through 3 May 2011 on a grass field near Delft University of Technology, Delft, the Netherlands (51–59′45.44″ N, 4–22′39.56″ E). The DTS instrument was a HALO unit (Sensornet, Elstree, UK) with a sampling interval of 2 m and a measurement interval of 20 s. The fiber optic cables used in this experiment consisted of single (simplex) multi-mode, bend-insensitive optical fibers, tightly packed, protected with Kevlar and a plastic jacket (AFL, South Carolina, USA).

A schematic drawing of the experimental setup is shown in Fig. 1. The cable consisted of four sections: one black with diameter 3.0 mm, one black with diameter 1.6 mm, one white with diameter 3.0 mm and one white with diameter 1.6 mm. Each section had a length of 190 m. Of each section, 150 m of fiber optic cable was held in open air, 1 m above the grass, with the use of pigtail fence posts. The sections were spliced together to enable continuous measurements.

The DTS was operated in a single-ended configuration. For calibration purposes, each section had 20 m of fiber cable coiled up in a thermally insulated water bath with warm water (average 27 °C) and cold water (average 14 °C). The setup of baths and splices was such that the cable from each stretch went directly through the two baths without first passing through a splice, thereby avoiding step losses within stretches. The signal was checked to ensure sufficient strength, especially towards the end of the cable. Bath temperatures were measured with the two PT100 thermometers that came with the HALO unit, which have a reported accuracy of 0.1 K. Calibration of the fiber optic cable was based on the method described by Hausner et al. (2011). For each measurement period, the bath temperatures from that same measurement period were used for the calibration.

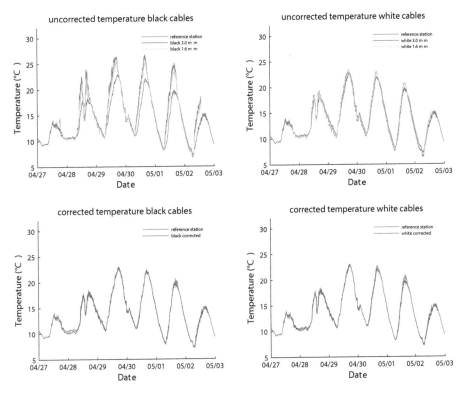

Figure 2. Top: comparison between the reference station and the uncorrected averaged temperature of the black (left) and white (right) cables in the measurement section. Bottom: comparison between the reference station and the corrected temperatures of the black (left) and white (right) cable.

For this analysis, 62 measurement points of each section hanging in the air were averaged to one time series. To improve further the signal to noise ratio, the DTS measurements were averaged over time using an integration time of 5 min. We assume that heat within the cross section of the cable is distributed equally at this timescale.

As a reference station, a HOBO weather station (Onset Computer Corporation, USA) with rain (mm), temperature ($^\circ$C), relative humidity (%) and incoming solar radiation (W m^{-2}) was installed next to the experimental setup. The temperature/RH sensor used was a 12 bit Temperature Smart Sensor (S-THB-M002) with a reported accuracy of $\pm 0.13\,^\circ$C. The reference station had a measurement interval of 5 min. The sensor was placed inside a HOBO RS1 radiation shield with only natural ventilation.

Wind velocity data measurements, to determine whether the assumption of forced convection was valid, were taken from the closest automated weather station of the Royal Netherlands Meteorological Institute. This station is situated at the airport of Rotterdam (51–57$'$33.66$''$ N, 4–26$'$32.66$''$ E), 6 km from the experiment location. There can be important differences between the wind measured in Rotterdam and the wind at our site, but these data only served as a check to see whether there were no periods without wind ($< 0.1\,\mathrm{m\,s^{-1}}$).

3 Results and discussion

Weather conditions from 27 to 30 April were partly cloudy with a daily maximum incoming solar radiation between 600 and 650 W m^{-2}. On 28 April there was a rain event with 2.4 mm of rain. Conditions from 1 to 3 May were clear and sunny with daily maximum incoming solar radiation between 650 and 750 W m^{-2}.

The dominant wind direction was north-east. The wind speed during daytime varied between 3 and 9 m s^{-1}, making the assumption of forced convection valid (Ar < 0.001). Note that forced convection will dominate in all but the most extreme natural conditions.

Figure 2 shows the average temperature measured by the black cables and the white cables. The uncorrected temperatures show a clear temperature rise during daytime due to solar heating. This effect is significantly larger for the black cables than for the white cables. During daytime, the black cables show a clear temperature rise due to solar heating. The correlation coefficient r^2 (see Fig. 3) between the uncorrected temperature of the black 3.0 and 1.6 mm cable with the reference station is respectively $r^2 = 0.77$ and $r^2 = 0.85$ and a RMSE of 2.40 and 1.80 $^\circ$C. The uncorrected temperatures of the white 3.0 and 1.6 mm cable show a correlation of respectively $r^2 = 0.97$ and $r^2 = 0.98$ and a RMSE of 0.74

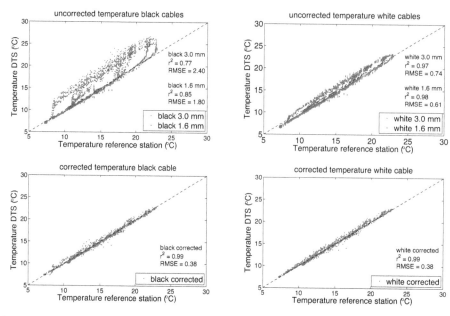

Figure 3. Top: correlations between the temperature measurements of the reference station and the uncorrected temperatures of the black (left) and white (right) table. Bottom: correlation between the temperature measurements of the reference station and the corrected temperatures of the black (left) and white (right) fiber optic cables.

and $0.61\,^{\circ}\mathrm{C}$. The corrected temperatures of both the white and black cable have an r^2 of 0.99 and a RMSE of $0.38\,^{\circ}\mathrm{C}$.

The standard deviations of individual measurements within the stretches were $0.29\,^{\circ}\mathrm{C}$ for $3.0\,\mathrm{mm}$ white, $0.28\,^{\circ}\mathrm{C}$ for $1.6\,\mathrm{mm}$ white, $0.39\,^{\circ}\mathrm{C}$ for $3.0\,\mathrm{mm}$ black, and $0.35\,^{\circ}\mathrm{C}$ for $1.6\,\mathrm{mm}$ black. The standard deviation of the average of the 62 points would then be between 0.04 and $0.05\,^{\circ}\mathrm{C}$. It should be noted that the accuracy of the instrument is about $0.02\,^{\circ}\mathrm{C}$ for a point measurement with perfect calibration and that similar setups have given field accuracies of $0.1\,^{\circ}\mathrm{C}$ (Hausner et al., 2011). So the measured variation is likely to be also caused by real temperature differences along the cable, caused by uneven heating/cooling.

4 Conclusions

Distributed temperature sensing (DTS) of atmospheric temperature profiles is hindered by solar heating, which may lead to significant deviations from the true air temperature. For atmospheric measurements with DTS we showed that it is possible to correct for solar heating and find a good estimation for the air temperature, by using cables with different diameters.

The corrected temperatures matched the temperature measurements of the reference station with a RMSE of $0.38\,^{\circ}\mathrm{C}$. The reference measurement took place without forced ventilation so the RMSE of the reference station could easily account for half the RMSE. The method used to calculate the air temperature is independent of the color of the cable. If it is not possible to apply different sizes of cable in a setup in

an atmospheric DTS application, the use of a thin white fiber optic cable is a reasonably good alternative.

The experiment started out with the question of whether corrected DTS air temperature measurements would have a reasonable accuracy to measure temperatures across urban landscapes. The results show that this is indeed the case and they were actually better than we expected at the onset ($1\,^{\circ}\mathrm{C}$). In hindsight, we would have installed a better reference temperature measurement device because we can not say whether the reference was truly better than the DTS-derived temperatures. Although we can not conclude this from the results, it may be possible to further reduce systematic errors, in which case this method would become valuable for vertical atmospheric soundings with DTS from balloons, quadcopters, or towers. White would be preferable over black, given the lower standard deviation within stretches.

Edited by: M. Hamilton

References

Ciocca, F., Lunati, I., Van de Giesen, N. C., and Parlange, M. B.: Heated optical fiber for distributed soil-moisture measurements: A lysimeter experiment, Vadose Zone J., 11, doi:10.2136/vzj2011.0199, 2012.

Curtis, A. and Kyle, P.: Geothermal point sources identified in a fumarolic ice cave on erebus volcano, antarctica using fiber optic distributed temperature sensing, Geophys. Res. Lett., 38, L16802, doi:10.1029/2011GL048272, 2011.

Dornstadter, M. and Aufleger, D.: The Prospect for Reservoirs in the 21st Century: Proceedings of the Tenth Conference of the BDS Held at the University of Wales, Bangor on 9–12 September 1998, Thomas Telford Publishing, 1998.

Freifeld, B. M., Finsterle, S., Onstott, T. C., Toole, P., and Pratt, L. M.: Ground surface temperature reconstructions: Using in situ estimates for thermal conductivity acquired with a fiber-optic distributed thermal perturbation sensor, Geophys. Res. Lett., 35, L14309, doi:10.1029/2008GL034762, 2008.

Hausner, M. B., Suárez, F., Glander, K. E., Van de Giesen, N., Selker, J. S., and Tyler, S. W.: Calibrating single-ended fiber-optic raman spectra distributed temperature sensing data, Sensors, 11, 10859–10879, doi:10.3390/s111110859, 2011.

Henderson, R. D., Day-Lewis, F. D., and Harvey C. F.: Investigation of aquifer-estuary interaction using wavelet analysis of fiber-optic temperature data, Geophys. Res. Lett., 36, L06403, doi:10.1029/2008GL036926, 2009.

Hoes, O., Schilperoort, R., Luxemburg, W., Clemens, F., and Van de Giesen, N.: Locating illicit connections in storm water sewers using fiber-optic distributed temperature sensing, Water Res., 43, 5187–5197, doi:10.1016/j.watres.2009.08.020, 2009.

Jansen, J., Stive, P., Van de Giesen, N., Tyler, S., Steele-Dunne, S., and Williamson, L.: Estimating soil heat flux using distributed temperature sensing, in: IAHS Publ. 343, 140–144, 2011.

Keller, C. A., Huwald, H., Vollmer, M. K., Wenger, A., Hill, M., Parlange, M. B., and Reimann, S.: Fiber optic distributed temperature sensing for the determination of the nocturnal atmospheric boundary layer height, Atmos. Meas. Tech., 4, 143–149, doi:10.5194/amt-4-143-2011, 2011.

Krause, S., Taylor, S. L., Weatherill, J., Haffenden, A., Levy, A., Cassidy, N. J., and Thomas, P. A.: Fibre-optic distributed temperature sensing for characterizing the impacts of vegetation coverage on thermal patterns in woodlands, Ecohydrology, 6, 754–764, doi:10.1002/eco.1296, 2013.

Lowry, C. S., Walker, J. F., Hunt, R. J., and Anderson, M. P.: Identifying spatial variability of groundwater discharge in a wetland stream using a distributed temperature sensor, Water Resour. Res., 43, W10408, doi:10.1029/2007WR006145, 2007.

Mamer, E. A. and Lowry, C. S.: Locating and quantifying spatially distributed groundwater-surface water interactions using temperature signals with paired fiber-optic cables, Water Resour. Res., 49, 7670–7680, doi:10.1002/2013WR014235, 2013.

Petrides, A. C., Huff, J., Arik, A., Van de Giesen, N., Kennedy, A. M., Thomas, C. K., and Selker, J. S.: Shade estimation over streams using distributed temperature sensing, Water Resour. Res., 47, W07601, doi:10.1029/2010WR009482, 2011.

Read, T., Bour, O., Bense, V., Le Borgne, T., Goderniaux, P., Klepikova, M., Hochreutener, R., Lavenant, N., and Boschero, V.: Characterizing groundwater flow and heat transport in fractured rock using fiber-optic distributed temperature sensing, Geophys. Res. Lett., 40, 2055–2059, doi:10.1002/grl.50397, 2013.

Sayde, C., Gregory, C., Gil-Rodriguez, M., Tufillaro, N., Tyler, S., Van de Giesen, N., English, M., Cuenca, R., and Selker, J. S.: Feasibility of soil moisture monitoring with heated fiber optics, Water Resourc. Res., 46, W06201, doi:10.1029/2009WR007846, 2010.

Selker, J., Van de Giesen, N., Westhoff, M., Luxemburg, W., and Parlange, M. B.: Fiber optics opens window on stream dynamics,

Geophys. Res. Lett., 33, L24401, doi:10.1029/2006GL027979, 2006.

Steele-Dunne, S. C., Rutten, M. M., Krzeminska, D. M., Hausner, M., Tyler, S. W., Selker, J., Bogaard, T. A., and Van de Giesen, N. C.: Feasibility of soil moisture estimation using passive distributed temperature sensing, Water Resour. Res., 46, W03534, doi:10.1029/2009WR008272, 2010.

Suárez, F., Aravena, J. E., Hausner, M. B., Childress, A. E., and Tyler, S. W.: Assessment of a vertical high-resolution distributed-temperature-sensing system in a shallow thermohaline environment, Hydrol. Earth Syst. Sci., 15, 1081–1093, doi:10.5194/hess-15-1081-2011, 2011.

Tanimola, F. and Hill, D.: Distributed fibre optic sensors for pipeline protection, J. Nat. Gas Sci. Eng., 1, 134–143, doi:10.1016/j.jngse.2009.08.002, 2009.

Thomas, C., Kennedy, A., Selker, J., Moretti, A., Schroth, M., Smoot, A., Tufillaro, N., and Zeeman, M.: High-resolution fibre-optic temperature sensing: A new tool to study the two-dimensional structure of atmospheric surface-layer flow, Bound.-Lay. Meteorol., 142, 177–192, 2012.

Tyler, S. W., Selker, J. S., Hausner, M. B., Hatch, C. E., Torgersen, T., Thodal, C. E., and Schladow, S. G.: Environmental temperature sensing using raman spectra dts fiber-optic methods, Water Resour. Res., 45, W00D23, doi:10.1029/2008WR007052, 2009.

Van de Giesen, N., Steele-Dunne, S. C., Jansen, J., Hoes, O., Hausner, M. B., Tyler, S., and Selker, J.: Double-Ended Calibration of Fiber-Optic Raman Spectra Distributed Temperature Sensing Data, Sensors, 12, 5471–5485, 2012.

Van Emmerik, T., Rimmer, A., Lechinsky, Y., Wenker, K., Nussboim, S., and Van de Giesen, N.: Measuring heat balance residual at lake surface using distributed temperature sensing, Limnol. Oceanogr-Meth., 11, 79–90, 2013.

Vercauteren, N., Huwald, H., Bou-Zeid, E., Selker, J. S., Lemmin, U., Parlange, M. B., and Lunati, I.: Evolution of superficial lake water temperature profile under diurnal radiative forcing, Water Resour. Res., 47, W09522, doi:10.1029/2011WR010529, 2011.

Vogt, T., Schneider, P., Hahn-Woernle, L., and Cirpka, O. A.: Estimation of seepage rates in a losing stream by means of fiber-optic high-resolution vertical temperature profiling, J. Hydrol., 380, 154–164, doi:10.1016/j.jhydrol.2009.10.033, 2010.

Westhoff, M. C., Savenije, H. H. G., Luxemburg, W. M. J., Stelling, G. S., van de Giesen, N. C., Selker, J. S., Pfister, L., and Uhlenbrook, S.: A distributed stream temperature model using high resolution temperature observations, Hydrol. Earth Syst. Sci., 11, 1469–1480, doi:10.5194/hess-11-1469-2007, 2007.

Westhoff, M. C., Gooseff, M. N., Bogaard, T. A., and Savenije, H. H. G.: Quantifying hyporheic exchange at high spatial resolution using natural temperature variations along a first-order stream, Water Resour. Res., 47, W10508, doi:10.1029/2010WR009767, 2011.

White, F.: Heat and Mass Transfer, Addison-Wesley series in mechanical engineering, Addison Wesley Publishing Company Incorporated, Reading, Massachusetts, USA, 718 pp., 1988.

WMO: WMO Guide to meteorological instruments and methods of observation, WMO-No. 8 (2008 edition, Updated in 2010), 2010.

Yilmaz, G. and Karlik, S. E.: A distributed optical fiber sensor for temperature detection in power cables, Sensor. Actuator. A-Phys., 125, 148–155, doi:10.1016/j.sna.2005.06.024, 2006.

Adaptive neuro-fuzzy inference system for temperature and humidity profile retrieval from microwave radiometer observations

K. Ramesh[1], **A. P. Kesarkar**[2], **J. Bhate**[2], **M. Venkat Ratnam**[2], **and A. Jayaraman**[2]

[1]Department of Computer Applications, Anna University, Regional Center, Tirunelveli, Tamil Nadu 627 005, India
[2]National Atmospheric Research Laboratory, Gadanki 517 112, Chittoor District, Andhra Pradesh, India

Correspondence to: A. P. Kesarkar (amit@narl.gov.in, amit.kesarkar@gmail.com)

Abstract. The retrieval of accurate profiles of temperature and water vapour is important for the study of atmospheric convection. Recent development in computational techniques motivated us to use adaptive techniques in the retrieval algorithms. In this work, we have used an adaptive neuro-fuzzy inference system (ANFIS) to retrieve profiles of temperature and humidity up to 10 km over the tropical station Gadanki (13.5° N, 79.2° E), India. ANFIS is trained by using observations of temperature and humidity measurements by co-located Meisei GPS radiosonde (henceforth referred to as radiosonde) and microwave brightness temperatures observed by radiometrics multichannel microwave radiometer MP3000 (MWR). ANFIS is trained by considering these observations during rainy and non-rainy days (ANFIS(RD + NRD)) and during non-rainy days only (ANFIS(NRD)). The comparison of ANFIS(RD + NRD) and ANFIS(NRD) profiles with independent radiosonde observations and profiles retrieved using multivariate linear regression (MVLR: RD + NRD and NRD) and artificial neural network (ANN) indicated that the errors in the ANFIS(RD + NRD) are less compared to other retrieval methods.

The Pearson product movement correlation coefficient (r) between retrieved and observed profiles is more than 92 % for temperature profiles for all techniques and more than 99 % for the ANFIS(RD + NRD) technique Therefore this new techniques is relatively better for the retrieval of temperature profiles. The comparison of bias, mean absolute error (MAE), RMSE and symmetric mean absolute percentage error (SMAPE) of retrieved temperature and relative humidity (RH) profiles using ANN and ANFIS also indicated that profiles retrieved using ANFIS(RD + NRD) are signif-

icantly better compared to the ANN technique. The analysis of profiles concludes that retrieved profiles using ANFIS techniques have improved the temperature retrievals substantially; however, the retrieval of RH by all techniques considered in this paper (ANN, MVLR and ANFIS) has limited success.

1 Introduction

Atmospheric convection plays an important role in the energy circulation of the atmosphere by transporting heat, momentum and moisture from the boundary layer to the free atmosphere. The vertical transport of these fluxes (heat, momentum and moisture) determines the evolution of multiscale convective phenomena such as thunderstorms and tornadoes (Lane and Moncrieff, 2010; Shaw and Lane, 2013). The temporal scale of these phenomena ranges from a few minutes to hours, and they are associated with disastrous effects that are of socioeconomic importance (Doswell III, 1985). Therefore, a continuous monitoring of the profiles of the atmosphere is important for their study. Conventionally, profiles of temperature and humidity are observed using radiosonde (GPS sonde hereafter referred to as radiosonde) measurements. However, it is difficult to study the evolution of convection using these observations due to their temporal resolution (frequency of vertical profiles). Further, these observations have a limited availability: operational radiosonde profiles are generally available at 00:00 and 12:00 UTC of every day as it is very expensive to launch radiosonde operationally at regular intervals of 1 h. Therefore, it is difficult to monitor the convective systems which evolve during the interval in between these launches. Moreover, the network of radiosonde observations is spatially coarse, and many times convection may not occur where the radiosonde is flying. Sometimes, updrafts and downdrafts present in the convection cause either spatial drift of the radiosonde or the bursting of the rubber balloon. On the other hand, space-based measurements of vertical profiles of the atmosphere using radio and microwave radars or radiometers on low Earth-orbiting satellites, sun synchronous satellites or geostationary satellites are useful to identify the convections, their movement and evolution. However, their revisit time/frequency of the observations and limited retrieval skill in the lower part of the atmosphere does not allow investigating the genesis and evolution of convection in most of the cases.

In this situation, multichannel microwave radiometers (MWRs) have evolved as powerful tools for monitoring the genesis and evolution of the convection over a station (Chan, 2009). MWR enables continuously monitoring microwave brightness temperatures, from which temperature, relative humidity and liquid water content can be derived. There are many studies targeting the retrieval of temperature and humidity profiles using MWR (Waters et al., 1975; Pandey and Kakar, 1983; Rodgers, 2000; Ware et al., 2003; Löhnert et al., 2004; Rose et al., 2005; Knupp et al., 2009; Matzler and Morland, 2009; Löhnert and Maier, 2012; Stähli et al., 2013; Xu et al., 2014). These investigations are aimed at determining temperature and water vapour soundings by observing radiated power at different microwave frequencies. Snider and Hazen (1998) described the observations of water vapour and cloud liquid based on MWR at frequencies of 20, 23, 31 and 90 GHz. D'Auria et al. (1998) used 19, 35 and 85 GHz frequency observations to study cloud properties and to generate a database of cloud genera useful for radiative-transfer modelling. Westwater et al. (1998) deployed a scanning MWR operating at a frequency of 5 mm (60 GHz) to study differences in boundary layer evolution over land and ocean. Their results showed the excellent agreement between atmospheric temperatures estimated by MWR and other measurements (meteorological towers and IR measurement). Ware et al. (2003) chose 12 microwave observation frequencies (22.035, 22.235, 23.835, 26.235, 30.0, 51.25, 52.28, 53.85, 54.94, 56.66, 57.29 and 58.8 GHz) to determine temperature, humidity and cloud liquid profiles. For these calculations the observed radiative power at different microwave frequencies converted into brightness temperatures using Plank's law. The profiles of temperature, relative humidity and liquid water content are retrieved using these brightness temperatures.

There are many retrieval algorithms proposed by previous investigators. Basili et al. (2001) developed a method to retrieve temperature profiles by microwave radiometry using a priori information on atmospheric spatial–temporal evolution. Bleisch et al. (2011) discussed the technique of the retrieval of water vapour profiles using MWR operating at a frequency of 22 GHz and its application to retrieve humidity profiles in the upper troposphere and lower-stratospheric (UTLS) region. Cimini et al. (2003) discussed the performance, calibration and achievable accuracy of a set of four MWRs operating in the 20–30 GHz band for the Atmospheric Radiation Program field experiments. They found that the brightness temperature measurements for two identical instruments differed less than 0.2 K over a period of 24 h. Binco et al. (2005) have demonstrated the synergistic use of microwave radiometer profiles and wind profiler radar to retrieve atmospheric humidity. They used wind profiler radar to estimate the potential refractivity gradient profiles and optimally combined them with MWR-estimated potential temperatures in order to fully retrieve the humidity gradient profile. Their results showed the significant improvement in the spatial vertical resolution of the atmospheric humidity profilers. Iassamen et al. (2009) used 12 frequencies of MWR to analyse the statistical distribution of tropospheric water vapour content in clear and cloudy conditions. They found that, vertically integrated water vapour content follows a Weibull distribution. Also, the vertical profiles of water vapour content during clear and cloudy conditions are well described by a function of temperature of the same form as the Clausius–Clapeyron equation. Haobo et al. (2011) proposed a retrieval method for temperature and humidity profiles based on principal-component analysis and stepwise regression.

It is found from these studies that MWR is becoming a robust tool for the monitoring of brightness temperatures and retrieving temperature and humidity profiles and hence the thermodynamic conditions of the atmosphere, which are very important for studying convective storms (Chan, 2009; Cimini et al., 2011). Güldner and Spänkuch (2001) discussed remote sensing of the thermodynamic state of the atmospheric boundary layer using ground-based microwave radiometer. Chan (2009) discussed the use of an MWR thermodynamic profile for the nowcasting of severe weather, such as a rainstorm, using a humidity profile and K index. They found that the accumulation of water vapour and the increase in the instability in the troposphere 1 h prior to occurrences of heavy rain are useful for its nowcasting. Therefore, MWR is becoming a useful tool for the nowcasting of intense convective weather due to high-frequency and accurate measurement of thermodynamical profiles. These profiles are very important for understanding the mesoscale processes and physical mechanisms involved in the preconditioning and triggering of small-scale convections such as thunderstorms and tornados. and also for understanding their temporal evolution. This understanding is very important for studying global energy transport. However, only limited efforts exist, especially over the tropical region because of the unavailability of high-frequency observations over this region.

Recent developments in the retrieval algorithms and computational techniques are adaptive and devise a model (Gaffard and Hewison, 2003) which improves the performance and accuracy of radiometer retrievals. Many nonlinear statistical/evolutionary algorithms are being developed for retrieving the profiles of the atmosphere using MWR (Solheim et al., 1998). These include the artificial neural network (ANN), Newtonian iteration of statistically retrieved profiles and Bayesian most probable retrieval. ANNs are widely used for different types of infrared and microwave-sounding instruments (Frate and Schiavon, 1998; Binco et al., 2005). Frate and Schiavon (1998) presented an inversion technique to retrieve profiles of temperature and water vapour using MWR. Their techniques combined a profile over a complete set of orthogonal functions with ANN, which performs the estimate of the coefficient of the expansion itself. Their analysis shows that this technique is flexible and robust. Ajil et al. (2010) used a new nonlinear technique ANFIS (adaptive neuro-fuzzy inference system) to improve the first guess using simulated infrared brightness temperatures for Geostationary Operational Environmental Satellite (GOES)-12 sounder channels. They found that the results of ANFIS retrieval are robust and reduce the root mean squared error by 20 % compared to regression fitting. They also argued that, as ANFIS uses a fuzzy-information system (FIS) for the classification of input, the classification of the training data set is not needed as it is required for regression techniques. In the present work, we have developed an ANFIS model-based retrieval of atmospheric parameters using MWR observations at NARL (National Atmospheric Research Laboratory), India. The objective of this algorithm development is to improve the accuracy of the retrieval of temperature and humidity profiles of MWR especially over the lower atmosphere.

The paper is organized as follows. Section 2 of this paper describes the details of data used for this study. The details of the method used and the ANFIS algorithm are described in Sect. 3. The experimental results are discussed in Sect. 4, and conclusions obtained from this work are presented in Sect. 5.

2 Microwave radiometer

The principal sources of atmospheric microwave emissions and absorptions are weak electric dipole rotational transition and magnetic dipole transitions of water vapour, oxygen and cloud liquid water (Westwater, 1993). Therefore, continuous monitoring of these thermal radiations has potential applications in meteorology and related sciences. MWRs are used for monitoring these radiations and are useful for continuous thermodynamical soundings (Ware et al., 2003). These MWRs are generally passive radiometers, continuously monitoring brightness temperature at various wavelengths in the microwave region of electromagnetic spectra. Ware et al. (2003) described the details of the MWR instrument, which is useful for temperature, water vapour and moisture sounding in clear and cloudy conditions. This instrument monitors the water vapour absorption line at 22 GHz to determine the water vapour profile as the magnitude of pressure broadening of water vapour absorption

line at this frequency decreases with height. This instrument monitors radiated power in a molecular oxygen absorption band around 60 GHz to determine temperature profiles and radiative power at selected frequencies of 22 to 59 GHz together to determine the liquid water profile. Cloud base height is estimated from zenith-infrared observations and retrieved temperature profiles. The MWR K band channels (22–30 GHz) are calibrated using tipping and V band channels (51–59 GHz) using a patented cryogenic black-body target. These calibrations are automatically transferred to a temperature-stabilized noise source. The internal mirror and azimuthal drive are used to point at any direction in the sky. The brightness temperatures are determined at various frequencies by using Plank's law and radiative-power observations (Han and Westwater, 2000; Ware et al., 2003). These brightness temperatures are used as input to the neural network for regression retrievals.

MWR is associated with the software (VIZMet-B)-enabled ANN retrieval algorithm for retrieving the profiles of temperature, relative humidity, liquid water content and vapour density. This ANN is a simple back-propagation neural network developed by Stuttgart University. The back-propagation algorithm is trained using microwave radiances observed by MWR as inputs and corresponding radiosonde observations as outputs. ANN generated weighing functions corresponding to different microwave frequencies as required by a radiative-transfer model, which, in turn, is useful for deriving the height profiles of temperatures and relative humidity. This MWR provides data with a vertical resolution of 50 m from surface up to a height of 500 m, 100 m from 500 m to 2 km and 250 m from 2 to 10 km. The further details of this MWR are available at the following website: http://www.radiometrics.com.

Gaffard and Hewison (2003), in their trial report on this radiometer (Radiometrics MP3000), have shown that the RMSE in the temperature profiles increases rapidly from 0.5 K at the surface to 1.5 K at 1 km and more slowly to 1.8 K at 5 km. According to Cimini et al. (2006, 2010), temperature and humidity retrieval accuracy is best near the surface and degrades with height; also, above 3 km, the retrieval accuracy and resolution degrade rapidly for all techniques. These studies used the observations reported without rain because the MWR cannot make any useful atmospheric observations during anything more than moderate rains. Thus, the major limitation of MWR is its performance degradation under heavy-precipitation conditions. Nevertheless, this instrument is believed to play an important role in investigating the thermodynamic condition of convection; however, the reliability and the performance can be enhanced by using better retrieval algorithms. Therefore, to improve/test the improvement of the accuracy of the retrieval of temperature and humidity profiles using MWR observations, we have developed the ANFIS system.

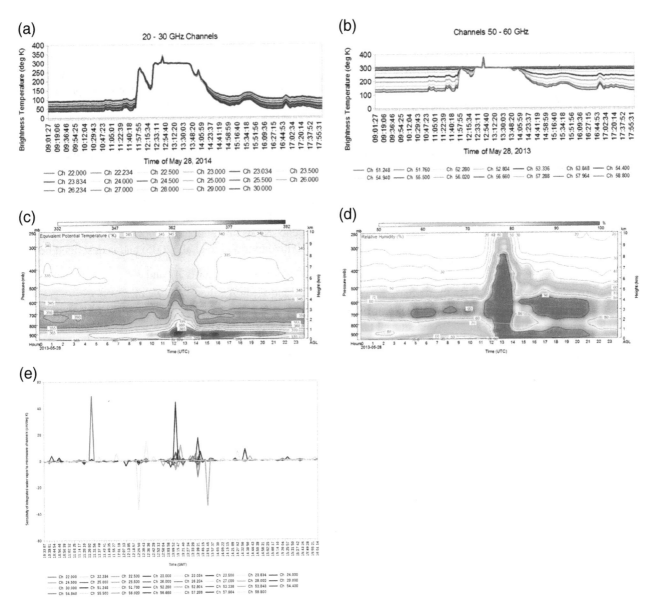

Figure 1. Sensitivity of 31 microwave brightness channels measured by MWR at **(a)** 20–30 GHz and **(b)** 50–60 GHz and the composite of vertical profiles of equivalent potential temperature **(c)** and relative humidity **(d)** retrieved during a convection event on 28 May 2013 over NARL Gadanki using MWR (ANN algorithm). The time resolution of these profiles is 4 min. **(e)** Sensitivities of integrated water vapour to brightness temperatures observed by MWR (rate of changes in integrated water vapour with respect to changes in brightness temperatures in each channel at a constant time span).

3 Data

At the National Atmospheric Research Laboratory, Gadanki (13.5° N, 79.2° E), India, MWR (MP3000-A manufactured by M/S Radiometrics, USA) is installed to study diurnal variations in convection and rainfall, for which an understanding of the genesis and further evolution of convection is very important. MWR at NARL has 31 channels in the microwave frequency range of 20–200 GHz (22 in K band and 14 in V band). For this study, we have used the observations from this MWR in zenith direction from 10 microwave channels,

viz. 22.234, 22.500, 23.034, 23.834, 25.000, 26.234, 28.000, 30.000, 57.964 and 58.800 GHz, to retrieve profiles of atmospheric temperature and relative humidity. These channels are selected based on the sensitivity of these channels during the occurrence of thunderstorms over the study site as shown in Fig. 1a and e. These figures show that these channels are sensitive to the advection of water vapour over this site (Fig. 1d) and its condensation during the period of 4 h prior to thunderstorm occurrence (Fig. 1c). Figure 1e shows the sensitivities of retrieved integrated water vapour content to microwave brightness temperatures at different frequencies.

For the formulation, training and validation of multivariate linear regression (MVLR), ANFIS and ANN systems, we have used the temperature and relative humidity observed by co-located GPS radiosonde (Meisei, Japan make, RS-01GII) measurements usually available almost every day at 12:00 UT (LT = UT + 05:30 h) at NARL Gadanki for the same period of training data set. Note that the Meisei radiosonde uses the temperature (relative humidity) sensors made with the thermistor (carbon humidity sensor), which measures the temperature (relative humidity) in the range of -900 to $+400\,^\circ$C (0–100 %) with an accuracy of 0.2 to 0.5 °C (2–5 %) (Basha and Ratnam, 2009).

In this work, we have used 122 days of MWR observations at the above-mentioned frequencies and radiosonde observations at 12:00 UTC during the period of June–September 2011. Out of 122 days (JJAS), 92 days are used for training the ANFISs (RD (rainy day) + NRD (non-rainy day)) and 30 days are used as an independent validation data set. The dates selected for independent validation are 24–30 June, 21–31 July, 26–31 August and 26–30 September 2011. ANFISs are trained using other 92-day observations excluding observations of the days selected for validation. Also, MVLR models are formulated using these 92-day observations and validated using the validation data set. The regular profiles of radiosondes are available every 12:00 UTC at the NARL site. Therefore, the ANFISs trained using 12:00 UTC observations. The ANFIS system would have been more robust if it had been trained using many radiosonde observations at regular intervals of each day. Unfortunately, obtaining periodic profiles of radiosondes at regular intervals of each day for long periods (monsoon months) to train the ANFIS system are not economically feasible. In this paper, for training and validation, we have sampled MWR data at 10 vertical locations at an interval of 1 km starting from 1 km. The vertical resolution of radiosonde data available for this study during the observational campaign is of 100 m resolution; therefore, the radiosonde observations of temperature and relative humidity at an altitude of within ±100 m of the target altitude are assumed at the sampled altitude.

4 Method

4.1 Fuzzy-information system

Fuzzy logic (FL) provides a simple way to arrive at a definite conclusion based upon vague, ambiguous, imprecise, noisy and missing input information (Priyono et al., 2005). Most of the FL models are empirically based, relying on an operator's experience rather than a technical understanding of the system. FL methods allow a number of inputs and generate a number of outputs. However, the generation of more inputs

and outputs will create more rules, and their interrelations make models more complex. To avoid the subjectivity in the operator's experience, Takagi and Sugeno (1985; TS85) proposed a mathematical tool to build a fuzzy model of the system. The TS85 system is based on the fuzzy partition of the input space into fuzzy subspace and on generating a linear relationship between each fuzzy subspace. Thus it forms a multidimensional fuzzy set in the product space of input variables to identify the premise of the fuzzy rule and then assigns linear consequents of each rule (Priyono et al., 2005). The identification of the fuzzy model can be improved using multidimensional reference fuzzy sets. The model is then structured into a set of IF–THEN statements. The Takagi–Sugeno–Kang fuzzy model composed of IF–THEN rules is described by Priyono et al. (2005) and is described below.

$$R_{(k)} : \text{If } x_1 \text{ is } A_k^1 \text{ and } x_2 \text{ is } A_k^2 \text{ and } \ldots$$
$$\text{and } x_m \text{ is } A_k^m \text{ then } y_k \text{ is } f_k(x), \quad (1)$$

where $f_k(x) = \alpha_k^0 + \alpha_k^1 x_1 + \alpha_k^2 x_2 + \ldots + \alpha_k^m x_m$ is a linear function and

- $k = 1 \ldots n$ denotes the node number

- y_k = output variables

- A_k^m = fuzzy sets (linguistic labels) associated with each node.

The above equation suggests that each fuzzy rule describes local linear behaviour. For any input $\hat{x} = (x_k^1, x_k^2 \cdots x_k^m)$ the inferred value of the Takagi–Sugeno–Kang (Takagi and Sugeno, 1985; Sugeno and Kang, 1988) fuzzy models is calculated as

$$\gamma = \frac{\sum\limits_{k=1}^{m} A_k(\hat{x})^*}{\sum\limits_{k=1}^{m} A_k(\hat{x})} = \frac{\sum\limits_{k=1}^{m} \tau_k \cdot f_k(\hat{x})}{\sum\limits_{k=1}^{m} \tau_k}, \quad (2)$$

where $A_k(\hat{x}) = \tau_k = A_k^1(x_k^1) \cdot A_k^2(x_k^2) \cdot \ldots \cdot A_k^m(x_k^m)$, τ_k is the level of firing of the kth rule for the current input $\hat{x}$. The model output is linear in weight but nonlinear in centre and standard deviation. The fuzzy clustering divides the input data space into fuzzy clusters, each representing one specific part of the system behaviour. There are several methodologies proposed for the clustering (Priyono et al., 2005). Chiu (1997, 1994) proposed the subtractive fuzzy-clustering method, and it is described in detailed by Priyono et al. (2005). We have used this method to build the fuzzy rules. This helps in reducing the number of rules and automatically determining the number of clusters (Chiu, 1994). The number of fuzzy rules varies depending on the total number of clusters (Chiu, 1997; Yager and Filev, 1994). Subtractive clustering finds the high-density region in the feature space (Jang, 1997; Jang et al., 2007). Subtractive clustering identifies the cluster centre in the data points with the following procedure:

1. Let N be the number of data points with n dimension vectors $x_k^i k = 1, 2, \ldots ni = 1, \ldots m$.

2. Density measure is calculated for each data point. A density measure at data point x_k^i is

$$D_k^i = \sum_{j=1}^{n} \exp\left(\frac{\|x_k^i - x_j^i\|^2}{(\frac{r_a}{2})} \right), \tag{3}$$

where r_a is the radius of the cluster. We have set its value to 0.3 in this analysis.

3. Based on the density measure, the data point with the highest density is selected as the first cluster centre x_{c1}, with a density measure of D_{c1}.

4. The density measure for each data point is revised by

$$D_k^i = D_k^i - D_{c1}^i \sum_{j=1}^{n} \exp\left(\frac{\|x_k^i - x_j^i\|}{(\frac{r_b}{2})^2} \right). \tag{4}$$

The constant r_b defines a neighbourhood to be reduced in density measure. To avoid repetition in the data points within the selected cluster, the data points within the cluster are discarded, and their absence is ensured in the next cluster. With the new feature space, a new high-density point is identified by the algorithm. This procedure is continued until all the data points are evaluated. Finally, the algorithm returns a set of clusters based on the Euclidean distance between the cluster centre and the data point (search radius).

4.2 ANFIS

ANFIS is a hybrid learning procedure which constructs an input–output mapping based on fuzzy if–then rules with appropriate member functions to generate the stipulated input–output pairs (Jang, 1993). ANFIS exploits the machine-learning potential of ANN and multi-valued logic of a fuzzy system in a single framework. Fuzzy logic is used for the classification of an input data set in different classes and forms the input to artificial neural networks. Then ANN is used to predict the output based on the training data sets. Thus, fuzzy logic controls the way of processing data by its classification to minimize the error in the neural network prediction (Tahmaseb and Hezarkhani, 2010). In recent decades, the ANFIS system has been used for many applications, such as turning tool-failure detection (Lo, 2002), quantitative structure activity relationships (Buyukbingol et al., 2007), drought forecasting (Bacanli et al., 2009), sea level prediction (Lin and Chang, 2008) and grade estimation (Tahmaseb and Hezarkhani, 2010). ANFIS caters to the need of complex real-world problems, which require intelligent systems that combine knowledge, techniques and methodologies from various sources.

In this work, the ANFIS is used with 10 predictors (brightness temperatures of 10 channels observed by MWR as mentioned above) as input to retrieve the temperature and humidity each at 10 sampled altitudes, i.e. to determine 20 outputs. This means output parameters are correlated in some fashion. We have used a Sugeno-type subtractive fuzzy clustering (Chiu, 1994) to reduce the number of predictors to decrease the training rule in FIS to make ANFIS more robust. The reduction in the number of rules automatically determines the number of clusters by assuming each data point as a potential cluster centre and creates clusters based on the density (Chiu, 1994). We found that subtractive-type clustering forms six rules for retrieval of temperature and seven rules for retrieval of RH with number of degrees of freedom equal to four and three respectively. The ANFIS model structure used in this work is shown in Fig. 2 and described in the next session.

4.3 ANFIS model structure

In this work, to profile the vertical distribution of temperature and relative humidity, a separate ANFIS model is developed for each level starting from 1 to 10 km with a vertical resolution of 1 km. Each ANFIS model in this work uses tier-3 architecture (Fig. 2) based on the fuzzy set if–then rules proposed by Takagi and Sugeno (1983). It comprises of five layers viz. input layer, input membership functions, rules, output membership functions and output. Layer 0 of this model passes the input to all membership functions by using the observed brightness temperature at 10 different microwave frequencies at each height level as mentioned earlier (i.e. $m = 10$). Layer 1 is known as the fuzzification layer, in which the input values of brightness temperatures (x) are normalized with a maximum equal to 1 and a minimum equal to 0. This layer uses Gaussian function for normalization. This process is termed fuzzification and each node k associated with the membership function O_k^1.

$$O_k^1 = \mu A_k(x_k) \tag{5}$$

As discussed earlier, x_k is the input, A_k are the linguistic labels associated with the membership function and μA_k is a Gaussian function written as

$$\mu A_k(x_k^i) = \exp\left[-\left(\frac{x_k^i - b_k}{a_k} \right)^2 \right], \tag{6}$$

where, $a_k b_k$ are model parameters determined quantitatively and responsible for variation in the shape of input membership functions.

Layer 2 multiplies input signals and sends product out. The node in layer 2 is the product of the degrees to which the inputs satisfy the membership functions, and it is found by

$$w_k = \Pi \mu A_k(x_k), k = 1, \ldots n. \tag{7}$$

Layer 3 is the normalization layer in which the ratio of each rule's firing strength is calculated with respect to the sum of

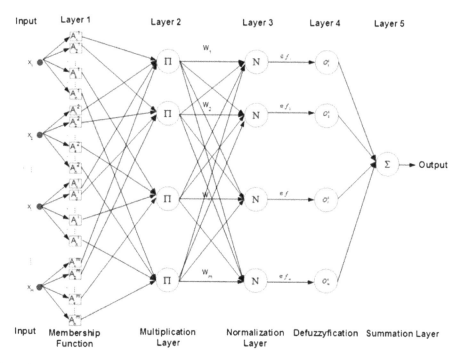

Figure 2. An ANFIS architecture for the Sugeno system used in this study (where $m = 10, n = 2, i = 1, 2$ and $k = 1, 2, \ldots, 10$).

the firing strengths of all the rules.

$$\bar{w}_k = \frac{w_k}{\sum\limits_{k=1}^{n} w_k} \tag{8}$$

The output of each node in layer 4 (defuzzification layer) is the weighted consequent value, and it is calculated by

$$O_i^4 = \bar{w} f_k = \bar{w}_k (\alpha_k^0 + \alpha_k^1 x_1 + \ldots + \alpha_k^m x_m), \tag{9}$$

where α_k^i is the parameter set.

Layer 5 is the summation layer, and its output is the sum of all the outputs of layer 4.

$$O_{5,1} = \sum_{k=1}^{n} \bar{w}_k f_k = \frac{\sum\limits_{k=1}^{n} w_k f_k}{\sum\limits_{k=1}^{n} w_k} \tag{10}$$

In this analysis, the FIS has been generated using the function genfis2 in MATLAB.

We have trained the ANFIS system in two ways: (1) by considering rainy days in the training data set and (2) by not considering rainy days in the training data sets. In this paper we have used ANFIS(NRD) to refer to ANFIS trained using microwave brightness temperature inputs only on non-rainy days and ANFIS(RD + NRD) to refer to ANFIS trained using microwave brightness temperature inputs on rainy and non-rainy days observed during the training period. The fitness of both the ANFIS and ANN models is tested as described below.

4.4 Multivariate linear regression

Multivariate linear regression (MVLR) is a classical linear statistical forecasting tool for understanding the relationship between a dependent variable and two or more independent variables. The multiple regression technique formulates a model to obtain estimates of the values of the dependent variable by fitting a linear equation to observed variables. Generally the form of the regression model is expressed as follows:

$$y_i = \beta_0 + \sum_{p=1}^{n} \beta_p x_{ip} + \epsilon_i \text{ where } i = 1, 2, \ldots m, \tag{11}$$

where y_i is a dependent variable which needs to be predicted (temperature and RH at different heights), x_{ip} is an independent variable (brightness temperatures measured by MWR at 10 different frequencies as mentioned above), β_p is a coefficient of linear regression which measures changes in y_i with respect to x_{ip}, ϵ_i is an error term representing the collective unobserved influence of any omitted variables, m is the number of in dependent variables, i.e. 10 in this paper, and n is the number of days used for training, i.e. 92 (total of 122 days of the months June to September 2011 – 30 days for independent verification) in this paper. Tables 1a, b and 2a, b list the values of β_p for temperature and RH profiles for MVLR(RD + NRD) and MVLR(NRD) respectively. In this paper, we have compared ANFIS(RD + NRD) and ANFIS(NRD) retrievals of profiles of temperature and RH with predicted profiles using MVLR. The results are discussed in the next section.

Adaptive neuro-fuzzy inference system for temperature and humidity profile retrieval from microwave...

13

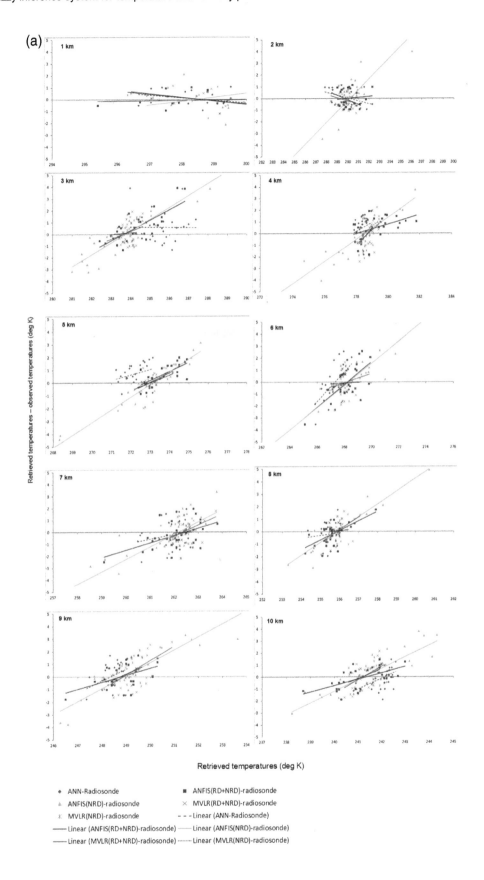

Figure 3.

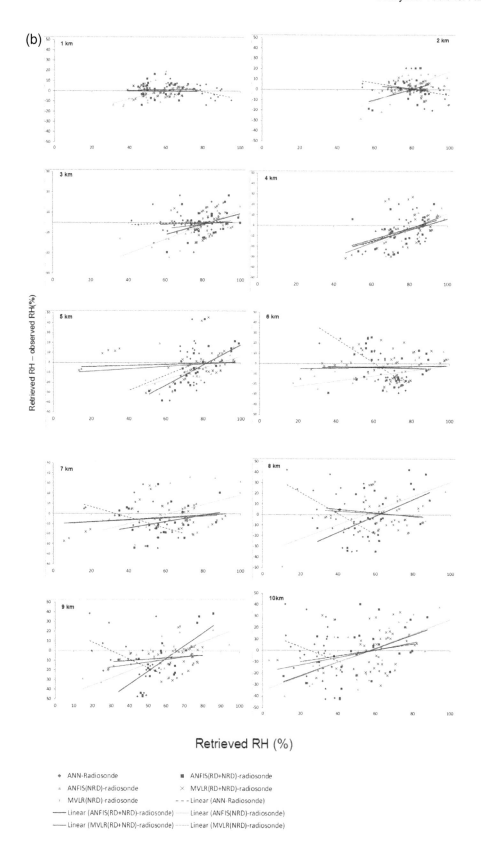

Figure 3.

Table 1a. Multivariate regression coefficients and intercepts for temperature approximation at different height levels. Coefficients are derived using rainy and non-rainy days data.

MVLR coefficients	Height (km)									
	1	2	3	4	5	6	7	8	9	10
Intercept	76.27	180.98	273.72	303.64	283.82	238.34	238.05	217.77	176.39	171.46
22.23	−0.16	0.16	0.26	0.25	0.52	0.29	0.57	1.00	0.81	0.60
22.5	0.54	0.06	−0.05	−0.57	−0.35	−0.34	−0.45	−0.69	−1.19	0.14
23.03	−0.60	−0.55	−0.39	−0.20	−0.86	−0.58	−0.60	−0.56	−0.18	−0.46
23.83	0.08	0.17	0.59	0.66	0.70	0.56	0.33	−0.02	1.22	−0.74
25	0.29	0.31	−0.53	0.14	0.20	0.19	0.70	0.74	−0.19	0.48
26.23	−0.11	0.37	−0.32	0.02	0.46	0.28	−0.46	−0.25	−0.84	−0.01
28	0.11	−0.34	0.77	−0.44	−0.91	−0.15	−0.15	−0.47	0.20	−0.05
30	−0.14	−0.12	−0.32	0.14	0.26	−0.22	0.06	0.28	0.19	0.04
57.96	0.13	0.00	0.00	0.11	−0.03	0.01	−0.10	−0.04	−0.18	0.03
58.8	0.61	0.36	0.04	−0.19	0.00	0.10	0.19	0.17	0.41	0.20

Table 1b. Same as Table 1a but for relative humidity.

MLR coefficients	Height (km)									
	1.00	2.00	3.00	4.00	5.00	6.00	7.00	8.00	9.00	10.00
Intercept	1364.58	545.24	−366.83	−532.92	63.44	720.22	196.00	856.69	688.20	811.07
22.23	−0.66	−3.92	−7.57	3.19	−4.87	7.27	6.43	−3.12	−5.39	1.46
22.50	−2.47	1.77	4.45	0.87	2.83	−9.07	9.03	12.32	20.06	8.33
23.03	2.22	4.15	3.49	−5.72	10.60	13.18	−5.21	3.23	−6.83	0.04
23.83	1.09	−1.36	−3.17	8.57	−3.62	−10.66	−12.02	−17.04	−9.83	−11.36
25.00	1.28	0.39	8.38	−0.29	2.32	4.99	−6.03	−2.70	−10.17	−6.24
26.23	3.13	−0.72	4.38	−8.46	−14.74	−15.01	7.93	5.35	15.11	2.82
28.00	−4.25	−1.34	−14.23	−3.42	−2.08	3.62	−3.07	6.36	1.22	11.45
30.00	0.08	1.20	5.06	5.81	9.47	5.59	2.97	−4.09	−3.79	−6.32
57.96	−0.95	0.21	2.43	2.03	0.88	−5.99	5.07	−0.61	1.74	2.53
58.80	−3.55	−1.93	−1.32	−0.54	−1.52	3.33	−5.85	−2.30	−4.04	−5.24

5 Results and discussions

5.1 MWR observations during convection

Figure 1a–e show the evolution of thunderstorm observed continuously (temporal resolution of temperature and relative humidity (RH) profiles: 4 min) by MWR on 28 May 2013. Figure 1a–b show the time series of different microwave channels at different frequencies between 20–30 and 50–60 GHz respectively. It can be seen from these figures that there is an increase in the magnitudes of brightness temperatures about 3 h prior to the occurrence of a thunderstorm. Therefore, the observed profiles of equivalent potential temperatures indicate preconditioning of the vertical column of the atmosphere to be conducive to the occurrence of thunderstorms about 3–4 h prior to their actual occurrence (Fig. 1c). The profile of relative humidity indicates the horizontal advection of moisture in a layer between 800–600 hPa and uplifting of moisture about 4 h prior to the occurrence of a thunderstorm. Therefore, MWR is found useful for in-vestigating the genesis and behaviour of the convection. The different microwave channel sensitivities to integrated water vapour content over the site of MWR are shown in Fig. 1e. As seen from this figure, different microwave frequencies are sensitive to changes in the water vapour content of the atmosphere. Figure 1a–e indicate that microwave brightness temperature observations can be used as a predictor for retrieving high-frequency profiles of relative humidity, and temperatures provided robust, reliable and accurate algorithms. In recent decades, ANFIS has been used for many applications, as mentioned above, because FIS trains back-propagation neural networks for different sets of input classification to generate robust results.

5.2 ANFIS training phase

The temperature and humidity profiles retrieved from ANFIS models for the training period are compared with the profile derived from GPS radiosonde observations. (Figure is not shown in the paper.) It is observed that during the

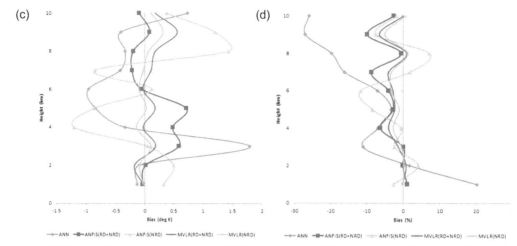

Figure 3. Scatter plot of difference between retrieved values of (**a**) temperature and (**b**) relative humidity using ANN, ANFIS(RD + NRD), ANFIS(NRD), MVLR(RD + NRD) or MVLR(NRD) techniques with radiosonde observations versus the retrieved values using these techniques respectively. Panels (**c**) and (**d**) are vertical distributions of bias in retrievals by different techniques for temperature and RH respectively.

Table 2a. Multivariate regression coefficients and intercepts for temperature approximation at different height levels. Coefficients are derived using only non-rainy days data.

MLR coefficients	Height (km)									
	1	2	3	4	5	6	7	8	9	10
Intercept	66.18	177.71	280.74	317.54	293.35	246.38	246.26	225.88	199.99	182.26
22.23	−0.06	0.43	−0.06	−0.13	−0.10	−0.10	0.02	0.55	0.09	0.52
22.5	0.34	−0.47	0.17	0.11	0.50	0.01	0.22	−0.09	−0.42	0.12
23.03	−0.46	−0.22	−0.17	−0.19	−0.95	−0.19	−0.57	−0.58	0.11	−0.51
23.83	0.06	0.08	0.29	0.18	0.50	0.04	0.24	−0.10	0.66	−0.46
25	0.18	0.37	−0.35	0.16	0.22	0.40	0.42	0.37	−0.27	0.30
26.23	0.01	0.28	−0.17	−0.05	0.37	0.26	−0.33	−0.13	−0.56	0.05
28	0.19	−0.18	0.73	0.00	−0.66	−0.25	0.13	0.03	0.61	0.16
30	−0.22	−0.22	−0.43	−0.09	0.13	−0.14	−0.12	−0.05	−0.19	−0.17
57.96	0.20	0.01	0.00	0.07	0.12	0.23	0.07	0.22	−0.02	0.05
58.8	0.57	0.36	0.00	−0.20	−0.18	−0.15	−0.01	−0.12	0.18	0.15

training period the values of the RMSE of temperature and relative humidity profiles are less than 0.01° C and 0.01 % respectively for all heights. The decrease in RMSE values regarding both RH and temperature retrieval are observed at heights of 2, 4 and 8 km for temperature retrieval. Similarly, for an RH profile there is a decrease in the RMSE values at 2, 6 and 9 km during the training period. It is seen that the number of radiosonde observations within 100 m of these sampled altitudes is higher compared to other altitudes. The decrease in the values of RMSE at this altitude may be due to the availability of relatively more samples for training. In general, it is found that, during the training phase, the ANFIS model shows a very good fit to radiosonde observations. Therefore, it is worth testing this model using an independent data set which is not considered for the training as discussed in Sect. 2.

5.3   Bias and scatter plots analysis

Figure 3a and b show the scatter plots between radiosonde observations and ANFIS(NRD), ANFIS(RD + NRD), MVLR(RD + NRD), MVLR(NRD) and ANN retrievals of temperature and relative humidity for different heights. The vertical profile of the bias in temperature and RH profiles is shown in Fig. 3c–d. It is seen from these figures that there is a significant reduction in the value of the bias for ANFIS(RD + NRD) and MVLR(RD + NRD) retrieval algorithms compared to MVLR(NRD), ANN and ANFIS(NRD) algorithms. However, it is seen from the analysis that ANN has relatively more systematic bias compared to ANFIS algorithms. More investigation in terms of the optimal amount of input data required for the appropriate classification using FIS and training of neural network is needed and is the aim of another publication.

Table 2b. Same as Table 2a but for relative humidity.

MLR coefficients	Height (km)									
	1	2	3	4	5	6	7	8	9	10
Intercept	1451.02	575.59	−557.29	−833.55	−304.23	920.46	660.87	973.55	370.93	517.24
22.23	−0.73	−4.98	−2.47	5.02	2.49	3.83	5.51	1.20	3.97	4.90
22.50	−2.34	4.47	−0.09	−3.84	−2.92	−1.98	8.46	7.76	13.71	4.98
23.03	2.31	1.58	2.01	−2.20	5.20	8.20	−2.99	2.69	−12.42	1.06
23.83	0.79	−0.14	−0.22	9.07	3.21	−7.25	−12.27	−16.59	−7.49	−17.23
25.00	1.67	−0.21	7.45	−3.01	−2.06	0.27	−5.19	−0.96	−8.32	2.01
26.23	2.02	0.01	0.59	−5.74	−14.80	−13.06	3.54	4.12	15.94	3.99
28.00	−3.25	−1.79	−12.13	−5.59	2.11	5.26	−1.03	5.09	−5.54	2.89
30.00	−0.10	1.22	5.59	6.87	6.82	4.53	3.75	−3.00	0.49	−2.35
57.96	−1.71	0.36	2.01	0.37	0.93	−3.58	1.63	−3.02	0.32	1.55
58.80	−3.06	−2.16	−0.27	2.09	−0.26	0.28	−3.92	−0.27	−1.53	−3.32

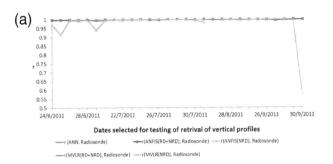

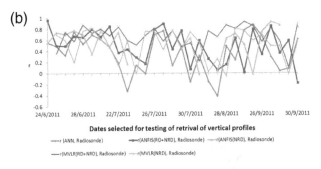

Figure 4. Pearson product movement correlation coefficient (r) between radiosonde (**a**) temperature and (**b**) humidity profiles with retrieved profiles using ANN, ANFIS(RD + NRD), ANFIS(NRD), MVLR(RD + NRD) and MVLR(NRD).

5.4 Correlation between retrieved and radiosonde profiles

The values of r calculated for the dates selected for the testing of retrieved profiles are shown in Fig. 4a and b. The r values for the temperature retrieval are more than 0.99 for ANN and ANFIS(RD + NRD) algorithms, and the value is relatively less for ANFIS(NRD) but better than 0.92. This indicates that these algorithms are successful in retrieving temperature profiles. It is also seen from Fig. 4a that the performance of ANFIS(RD + NRD) for temperature retrieval is

slightly better compared to the other two algorithms. Therefore it may be stated that the retrieval of temperature profiles using ANFIS(RD + NRD) is more reliable and can be used for the investigation of the physical mechanism associated with tropical convective systems. However, the retrieval of RH is also very important for investigating different microphysical processes responsible for the convection. Figure 4b shows the values of r for RH retrieval. One of the limitations of radiosonde observations is that the radiosonde drifts far away due to heavy winds during dynamical weather conditions when, generally, the atmosphere is moist and cloudy. Therefore, the data set of RH may not represent true measurements above the region zenith of the MWR as RH has more spatial variability than temperature. Also, there is limited information content in the brightness temperatures for the vertical distribution of moisture. Therefore, it is difficult to correlate the RH-retrieved profiles with that observed with radiosonde measurements. Nevertheless, the values of r are more than 60 % for about 18, 13 and 9 cases out of 29 cases for the ANFIS(RD + NRD), ANFIS(NRD) and ANN algorithms. For the rest of the cases, the values of r are less than 60 %. In the case of the ANN(ANFIS) retrieval of RH, it is found that 4 (1) case(s) out of 29 cases are negatively correlated with the radiosonde measurements. Thus, we found that the retrieval of RH using ANFIS(RD + NRD) is comparatively better than that of other two algorithms. However, we believe that a detailed investigation is required to understand and improve the correlation between RH radiosonde profiles and retrieved profiles, especially in the cloudy atmosphere or convectively efficient environment. It is worth investigating the impact of clouds on MWR brightness temperatures and consequently the retrieval of the humidity profile. This requires understanding the environmental dependence of the brightness temperatures measured by MWR. The adaptive virtue of ANFIS makes them suitable for further improvement of the retrieval technique presented in this paper, with the above-mentioned considerations. However, we strongly

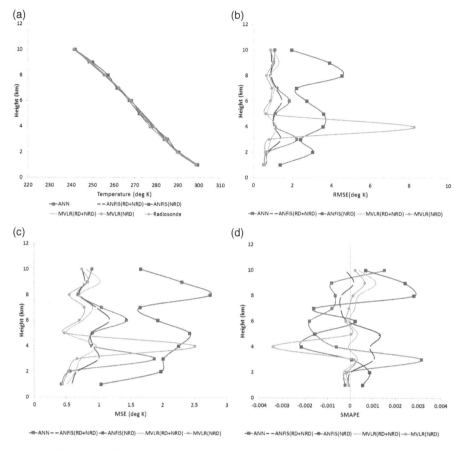

Figure 5. Comparison of vertical profiles of **(a)** temperatures observed by radiosonde and temperature profiles retrieved from ANN AN-FIS(RD + NRD), ANFIS(NRD), MVLR(RD + NRD) and MVLR(NRD) and **(b)** RMSE, **(c)** MAE and **(d)** SMAPE of retrieved profiles using ANN ANFIS(RD + NRD), ANFIS(NRD), MVLR(RD + NRD) and MVLR(NRD) with respect to temperature profiles observed by radiosonde.

feel that more systematic investigation is required to understand it, and we think that it should be addressed in another publication rather than in this paper.

5.5 Error analysis of retrieved temperature profiles

Figure 5a–d show the mean vertical profiles obtained by radiosonde profiles and retrieved from ANFIS(RD + NRD), ANFIS(NRD), MVLR(RD + NRD), MVLR(NRD) and ANN techniques. As mentioned in the previous section, it is seen from Fig. 5a that the mean (30 hypothesis testing days) observed and retrieved profiles overlap and have relatively very less errors. The RMSE for the verification data set is less than $0.7 \degree$C up to 2 km and shows a slight increase of 1 to $2.3 \degree$C at higher heights (Fig. 5b). The average error is $1.08 \degree$C. The profile of RMSE shows a small warm bias in the retrieved values of temperatures using the ANFIS(RD + NRD) model. However, ANFIS(RD + NRD) shows a significant reduction in bias and relatively better performance as compared to other two algorithms. The mean absolute error (MAE) for the test data set follows the qualitative trend of RMSE but is slightly less in magnitude.

The ANFIS(RD + NRD) algorithm has relatively less MAE. The behaviour of the symmetric mean absolute percentage error (SMAPE) (Fig. 5d) suggests that ANFIS(NRD) considers relatively more variation in temperature compared to the ANFIS(RD + NRD) and ANN algorithms and has a positive bias below 3 km and above 6 km and a negative bias in between 3 and 6 km.

Venkat Ratnam et al. (2013) have compared GPS radiosonde profiles with retrieved profiles using the ANN algorithm available with MWR (ANN–MWR). Their results showed that the warm (cold) bias between radiosonde and MWR in temperature is clearly observed below (above) 3–4 km depending upon the time. Madhulatha et al. (2013) have studied the mean profiles for temperature and vapour density and the difference between temperatures and vapour density along with standard deviations derived from ANN–MWR and a GPS radiosonde for the period June through December 2011. They found a very close agreement in temperature profiles between MWR and GPS radiosonde. Their results show differences in retrieved profiles with an ANN–MWR cold bias of about $2 \degree$C up to 4 km and a warm bias of about

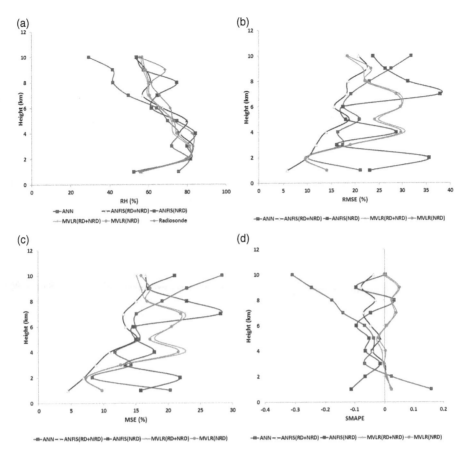

Figure 6. Comparison of vertical profiles of **(a)** RH observed by radiosonde and temperature profiles retrieved from ANN AN-FIS(RD + NRD), ANFIS(NRD), MVLR(RD + NRD) and MVLR(NRD) and **(b)** RMSE, **(c)** MAE and **(d)** SMAPE of retrieved profiles using ANN, ANFIS(RD + NRD), ANFIS(NRD), MVLR(RD + NRD) and MVLR(NRD) with respect to relative humidity profiles observed by radiosonde.

2 °C above 4 km. As seen from Fig. 5b, the ANFIS method is successful in reducing this bias with the average RMSE of 1.08.

5.6 Error analysis of retrieved humidity profiles

Figure 6a–d show the mean profile of retrieved relative humidity using ANFIS, ANN or MVLR models and observed brightness temperatures. The figure shows that the profile retrieved using the ANFIS(RD + NRD) model is qualitatively better compared to that using the ANN model. It is seen from Fig. 6b that the RMSE of retrieved relative humidity averaged over the training data set is less than 0.01 % throughout the profile. However, the values of RMSE of the testing data set for ANFIS, vary significantly (5–20 %) with respect to height. At 1 km, the value of RMSE is 4.87 %, at 2 km it is 6.19 % and it gradually increases towards higher heights up to a maximum of 23.89 % at 8 km. It is seen from Fig. 6b that ANFIS(RD + NRD) shows better performance than ANN in retrieving relative humidity. The variation of MAE is more or less consistent with the behaviour of RMSE. The behaviour of SMAPE with height shows that the AN-

FIS model takes into account more variability compared to ANN models but has a more negative bias at higher heights. The study by Venkat Ratnam et al. (2013) also indicated a large wet (dry) bias of 6–8 g kg^{-1} in the specific humidity below (above, except around 5–6 km) 2–3 km between the radiosonde and ANN algorithm.

6   Conclusions

In this work, we have presented a formulation of the AN-FIS model for the retrieval of atmospheric profile temperature and humidity using brightness temperatures observed at different microwave frequencies mentioned above by MWR. The ANFIS models are trained by considering rainy and non-rainy days together (ANFIS(RD + NRD)) and also only for non-rainy days (ANFIS(NRD)). In this work we found that ANFIS(RD + NRD) is more suitable for retrieving vertical profiles of the atmosphere by observing the power received on the ground due to different emissions at different microwave frequencies. Our results indicated that the performance of the ANFIS(RD + NRD) model is better than the

ANN back-propagation algorithm in retrieving profiles of both temperature and RH. The retrieved temperature profiles are relatively closer to the observations by radiosonde. However, an improvement is needed in the retrieval of relative humidity to reduce relatively large error at higher heights. For this purpose, a detailed investigation is required to be carried out to understand the behaviour of the brightness temperatures in a cloudy atmosphere and its impact on the weighting functions of MWR and the retrieval of vertical profiles using the ANFIS method. The development of robust algorithms for the retrieval of temperature and relative humidity using the new method ANFIS, especially during complex environmental conditions, will lead to MWR as a novel tool to investigate the physical mechanisms associated with small-scale convections.

Acknowledgements. The authors are thankful to V. Sundareswaran, Regional Director, Anna University, Regional Centre, Tirunelveli, India, for his continuous encouragement and support during this work. Also, thanks are due to A. Kiran Kumar, NARL, Gadanki, India, for his technical support during this work.

Edited by: M. Nicolls

References

Ajil, K. S., Thapliyal, P. K., Shukla, M. V., Pal, P. K., Joshi, P. C., and Navalgund, R. R.: A New Technique for Temperature and Humidity Profile Retrieval From Infrared-Sounder Observations Using the Adaptive Neuro-Fuzzy Inference System, IEEE T. Geosci. Remot., 48, 1650–1659, April, 2010.

Bacanli, U. G., Firat, M., and Dikbas, F.: Adaptive Neuro-Fuzzy Inference System for drought forecasting, Stoch. Env. Res. Risk A., 23, 1143–1154, 2009.

Basha, G. and Ratnam, M. V.: Identification of atmospheric boundary layer height over a tropical station using high resolution radiosonde reflectivity profiles: Comparison with GPS radio occultation measurements, J. Geophys. Res., 114, D16101, doi:10.1029/2008JD011692, 2009.

Basili, P., Bonafoni, S., Ciotti, P., Marzano, F. S., d'Auria, G., and Pierdicca, N.: Retrieving atmospheric temperature profiles by microwave radiometry using a priori information on atmospheric spatial-temporal evolution, IEEE T. Geosci Remote, 39, 1896–1905, 2001.

Binco, L., Cimini, D., Marzano, F. S., and Ware, R.: Combining Microwave Radiometer and Wind Profiler Radar Measurements for High-Resolution Atmospheric Humidity Profiling, J. Atmos. Ocean. Tech., 22, 949–965, doi:10.1175/JTECH1771.1, 2005.

Bleisch, R., Kämpfer, N., and Haefele, A.: Retrieval of tropospheric water vapour by using spectra of a 22 GHz radiometer, Atmos. Meas. Tech., 4, 1891–1903, doi:10.5194/amt-4-1891-2011, 2011.

Buyukbingol, E., Sisman, A., Akyildiz, M., Alparslan, F. N., and Adejare, A.: Adaptive neuro-fuzzy inference system (ANFIS): A new approach to predictive modelling in QSAR applications: A study of neuro-fuzzy modelling of PCP-based NMDA receptor antagonists, Bioorgan. Med. Chem., 15, 4265–4282, 2007.

Chan, P. W.: Performance and application of a multi-wavelength, ground-based microwave radiometer in intense convective weather, Meteorol. Z., 18, 253–265, 2009.

Chiu, S. L.: Fuzzy model identification based on cluster estimation, J. Intell. Fuzzy Syst., 2, 267–78, 1994.

Chiu, S. L.: Fuzzy Information Engineering: A Guided Tour of Applications, edited by: Dubois, D., Prade, H., and Yager, R., John Wiley & Sons, 1997.

Cimini, D., Westwater, E. R., Han, Y., and Keihm, S. J.: Accuracy of ground-based microwave radiometer and balloon-borne measurements during WVIOP2000 field experiment, IEEE T. Geosci. Remote, 41, 2605–2615, 2003.

Cimini, D., Hewison, T. J., Martin, L., Güldner, J., Gaffard, C., and Marzano, F.: Temperature and humidity profile retrievals from ground based microwave radiometers during TUC, Meteorol. Z., 15, 45–56, 2006.

Cimini, D., Westwater, E. R., and Gasiewski, A. J.: Temperature and humidity profiling in the Arctic using milli meter-wave radiometry and 1-DVAR, IEEE T. Geosci. Remote, 48, 1381–1388, doi:10.1109/TGRS.2009.2030500, 2010.

Cimini, D., Campos, E., Ware, R., Albers, S., Giuliani, G., Oreamuno, J., Joe, P., Koch, S., Cober, S., and Westwater, E.: Thermodynamic atmospheric profiling during the 2010 Winter Olympics using ground-based microwave radiometry, IEEE T. Geosci. Remote, 49, 4959–4969, doi:10.1109/TGRS.2011.2154337, 2011.

d'Auria, G., Marzano, F. S., Pierdicca, N., Pinna Nossai, R., Basili, P., and Ciotti, P.: Remotely sensing cloud properties from microwave radiometric observations by using a modelled cloud database, Radio Sci., 33, 369–392, 1998.

Doswell III, C. A.: The Operational Meteorology of Convective Weather. Vol. II: Storm-scale Analysis, NOAA Tech. Memo. ERL ESG-15, available from the author at National Severe Storms Lab., 1313 Halley Circle, Norman, OK, 73069, 1985.

Frate, D. F. and Schiavon, G.: A combined natural orthogonal functions/neural network technique for the radiometric estimation of atmospheric profiles, Radio Sci., 33, 405–410, doi:10.1029/97RS02219, 1998.

Gaffard, C. and Hewison, T.: Radiometrics MP3000 Microwave Radiometer, Trial Report Version 1.0, 2003.

Güldner, J. and Spänkuch, D.: Remote sensing of the thermodynamic state of the atmospheric boundary layer by ground-based microwave radiometry, J. Atmos. Ocean. Tech., 18, 925–933, 2001.

Han, Y. and Westwater, E. R.: Analysis and improvement of tipping calibration for ground-based microwave radiometers, IEEE T. Geosci. Remote, 38, 1260–1275, 2000.

Haobo, T., Mao, J., Chen, H., Chan, P. W., Wu, D., Li, F., and Deng, T.: A Study of a Retrieval Method for Temperature and Humidity Profiles from Microwave Radiometer Observations Based on Principal Component Analysis and Stepwise Regression, J. Atmos. Ocean. Tech., 28, 378–389, doi:10.1175/2010JTECHA1479.1, 2011.

Iassamen, A., Sauvageot, H., Jeannin, N., and Ameur, S.: Distribution of Tropospheric Water Vapour in Clear and Cloudy Conditions from Microwave Radiometric Profiling, J. Appl. Meteorol. Clim., 48, 600–615, 2009.

Jang, J. R.: ANFIS: Adaptive-Network-Based Fuzzy Inference System, IEEE T. Syst. Man Cybr., 23, 665–685, 1993.

Jang, J.-S. R., Sun, C. T., and Mizutani, E.: Neuro–Fuzzy and Soft Computing. Upper Saddle River, NJ: Prentice-Hall, ser. MATLAB Curriculum Series, 1997.

Jang, J. S. R., Sun, C.-T., and Mizutani, E.: Neuro-Fuzzy and soft computing: A computational approach to learning and machine Intelligence, Pearson Education, 2007.

Knupp, K. R., Coleman, T., Phillips, D., Ware, R., Cimini, D., Vandenberghe, F., Vivekanandan, J., and Westwater, E.: Ground-Based Passive Microwave Profiling during Dynamic Weather Conditions, J. Atmos. Ocean. Tech., 26, 1057–1073, 2009.

Lane, T. P. and Moncrieff, M. W.: Characterization of Momentum Transport Associated with Organized Moist Convection and Gravity Waves, J. Atmos. Sci., 67, 3208–3225, doi:10.1175/2010JAS3418.1, 2010.

Lin, L.-C. and Chang, H.-K.: An Adaptive Neuro-Fuzzy Inference System for Sea Level Prediction Considering Tide-Generating Forces and Oceanic Thermal Expansion, Terr. Atmos. Ocean. Sci., 19, 163–172, 2008.

Lo, S.-P.: The Application of an ANFIS and Grey System Method in Turning Tool-Failure Detection, Int. J. Adv. Manuf. Tech., 19, 564–572, 2002.

Löhnert, U., Crewell, S., and Simmer, C.: An integrated approach toward retrieving physically consistent profiles of temperature, humidity, and cloud liquid water, J. Appl. Meteorol., 43, 1295–1307, 2004.

Löhnert, U. and Maier, O.: Operational profiling of temperature using ground-based microwave radiometry at Payerne: prospects and challenges, Atmos. Meas. Tech., 5, 1121–1134, doi:10.5194/amt-5-1121-2012, 2012.

Madhulatha, A., Rajeevan, M., Ratnam, M. V., Bhate, J., and Naidu, C. V.: Now casting severe convective activity over south-east India using ground-based microwave radiometer observations, J. Geophy. Res., 118, 1–13, 2013.

Matzler, C. and Morland, J.: Refined physical retrieval of integrated water vapour and cloud liquid for microwave radiometer data, IEEE T. Geosci. Remote, 47, 1585–1594, 2009.

Pandey, P. C. and Kakar, R. K.: A Two Step Linear Statistical Technique Using Leaps And Bounds Procedure for Retrieval of Geophysical Parameters from Microwave Radiometric Data, IEEE T. Geosci. Remote, 21, 208–214, 1983.

Priyono, A., Ridwan, M., Allias, A. J., Atiq, R., Rahmat, O. K., Hassan, A., and Ali, M. A. M.: Generation of fuzzy rules with subtractive clustering, J. Teknologi, 43, 143–153, 2005.

Rodgers, C. D.: Inverse methods for atmospheric sounding: Theory and Practice, Series on Atmospheric, Oceanic and Planetary Physics, Vol. 2, Singapole, World Scentic, 2000.

Rose, T., Crewell, S., Löhnert, U., and Simmer, C.: A network suitable microwave radiometer for operational monitoring of the cloudy atmosphere, Atmos. Res., 75, 183–200, 2005.

Shaw, T. A. and Lane, T. P.: Toward an Understanding of Vertical Momentum Transports in Cloud-System-Resolving Model Simulations of Multi scale Tropical Convection, J. Atmos. Sci., 70, 3231–3247, 2013.

Solheim, F., Godwin, J. R., Westwater, E., Han, Y., Keihm, S. J., Marsh, K., and Ware, R.: Radiometric profiling of temperature, water vapour and cloud liquid water using various inversion methods, Radio Sci., 33, 393–404, 1998.

Snider, J. B. and Hazen, D. A.: Surface-based radiometric observations of water vapour and cloud liquid in the temperate zone and in the tropics, Radio Sci., 33, 421–432, 1998.

Stähli, O., Murk, A., Kämpfer, N., Mätzler, C., and Eriksson, P.: Microwave radiometer to retrieve temperature profiles from the surface to the stratopause, Atmos. Meas. Tech., 6, 2477–2494, doi:10.5194/amt-6-2477-2013, 2013.

Sugeno, M. and Kang, G. T.: Structure identification of fuzzy model, Fuzzy Sets Syst., 28, 15–33, 1988.

Takagi, T. and Sugeno, M.: Fuzzy identification of systems and its applications to modeling and control, IEEE Trans. Syst., Man, Cybern., 15, 116–132, 1985.

Tahmasebi, P. and Hezarkhani, A.: Application of Adaptive Neuro-Fuzzy Inference System for Grade Estimation; Case Study, Sarcheshmeh Porphyry Copper Deposit, Kerman, Iran, Australian Journal of Basic and Applied Sciences, 4, 408–420, 2010.

Takagi, T. and Sugeno, M.: Derivation of fuzzy control rules from human operator's control action, in: Proc. IFAC Symp. Fuzzy inform, Knowledge Representation and Decision Analysis, 55–60, 1983.

Venkat Ratnam, M. V., Durga Santhi, Y., Rajeevan, M., and Rao, S. V. B.: Diurnal variability of stability indices observed using radiosonde observations over a tropical station: Comparison with microwave radiometer measurements, Atmos. Res., 124, 21–33, 2013.

Ware, R., Carpenter, R., Güldner, J., Liljegren, J., Nehrkorn, T., Solheim, F., and Vandenberghe, F.: A multichannel radiometric profiler of temperature, humidity, and cloud liquid, Radio Sci., 38, 8079, doi:10.1029/2002RS002856, 2003.

Waters, J. W., Kunzi, K. F., Pettyjohn, R. L., Poon, R. K. L., and Staelin, D. H.: Remote Sensing of Atmospheric Temperature Profiles with the Nimbus 5 Microwave Spectrometer, J. Atmos. Sci., 32, 1953–1969, 10.1175/1520-0469(1975)032<1953:RSOATP>2.0.CO;2, 1975.

Westwater, E. R.: Ground-based microwave remote sensing of meteorological variables, in: Atmospheric Remote Sensing by Microwave Radiometry, J. Wiley Sons, Inc., edited by: Michael, A. J., Chapter 4, 145–213, 1993.

Westwater, Ed. R., Han, Y., Irisov, V. G., Leuskiy, V. Y., Trokhimovski, Y. G., Fairall, C. W., and Jessup, A. T.: Sea-air and boundary layer temperatures measured by a scanning 5-mm wavelength radiometer: Recent results, Radio Sci., 33, 291–302, 1998.

Xu, G., Ware, R., Zhang, W., Feng, G., Liao, K., and Liu, Y.: Effect of off-zenith observation on reducing the impact of precipitation on ground-based microwave radiometer measurement accuracy in Wuhan, Atmos. Res., 140–141, 85–94, 2014.

Yager, R. and Filev, D.: Generation of Fuzzy Rules by Mountain Clustering, J. Intell. Fuzzy Syst., 2, 209–219, 1994.

3

Multi-wavelength optical measurement to enhance thermal/optical analysis for carbonaceous aerosol

L.-W. A. Chen[1,2,3], J. C. Chow[2,3], X. L. Wang[2], J. A. Robles[2], B. J. Sumlin[2], D. H. Lowenthal[2], R. Zimmermann[4], and J. G. Watson[2,3]

[1]Department of Environmental and Occupational Health, University of Nevada, Las Vegas, Nevada 89154, USA
[2]Division of Atmospheric Sciences, Desert Research Institute, Reno, Nevada 89512, USA
[3]Key Laboratory of Aerosol Science & Technology, SKLLQG, Institute of Earth Environment, Chinese Academy of Sciences, Xi'an, China
[4]Joint Mass Spectrometry Centre, Chair of Analytical Chemistry, Institute of Chemistry, University of Rostock, Rostock, Germany

Correspondence to: L.-W. A. Chen (antony.chen@unlv.edu)

Abstract. A thermal/optical carbon analyzer equipped with seven-wavelength light source/detector (405–980 nm) for monitoring spectral reflectance (R) and transmittance (T) of filter samples allowed "thermal spectral analysis (TSA)" and wavelength (λ)-dependent organic-carbon (OC)–elemental-carbon (EC) measurements. Optical sensing was calibrated with transfer standards traceable to absolute R and T measurements, adjusted for loading effects to report spectral light absorption (as absorption optical depth ($\tau_{a,\lambda}$)), and verified using diesel exhaust samples. Tests on ambient and source samples show OC and EC concentrations equivalent to those from conventional carbon analysis when based on the same wavelength (~ 635 nm) for pyrolysis adjustment. TSA provides additional information that evaluates black-carbon (BC) and brown-carbon (BrC) contributions and their optical properties in the near infrared to the near ultraviolet parts of the solar spectrum. The enhanced carbon analyzer can add value to current aerosol monitoring programs and provide insight into more accurate OC and EC measurements for climate, visibility, or health studies.

1 Introduction

Thermal/optical analysis (TOA) quantifies particulate-matter (PM) organic carbon (OC) and elemental carbon (EC) collected on quartz-fiber filters (Watson et al., 2005; Cao et al., 2007; Bougiatioti et al., 2013). TOA based on the IMPROVE_A protocol (Chow et al., 2007a, 2011) has determined OC and EC concentrations in tens of thousands of samples each year from long-term chemical speciation networks operated in the US (IMPROVE, 2014; U.S.EPA, 2014), Canada (Dabek-Zlotorzynska et al., 2011), and China (Zhang et al., 2012). IMPROVE_A specifies stepped heating up to 580 °C in an inert helium (He) atmosphere (> 99.99 % purity), where most organic compounds are either evaporated or decomposed (Chow et al., 1993), followed by a second stage of stepped heating to 840 °C in 98 % He / 2 % O_2 to remove EC on the filter. Since some of the OC is converted to EC through pyrolysis in pure He, as evidenced by darkening of the filter, IMPROVE_A also specifies a reflectance pyrolysis adjustment. Reflected light at wavelength $\lambda = 633$ nm is monitored throughout the heating (Huntzicker et al., 1982). OC and EC are defined as carbon evolved before and after the filter reflectance (R) returns to its initial level, respectively.

In addition to reflectance, some TOA protocols use transmitted light (T) to monitor the pyrolysis (Birch and Cary, 1996; Turpin et al., 1990). ECs based on R or T splits are referred to as ECR and ECT, respectively. In addition to particle deposits, adsorbed organic vapors within the quartz-fiber filter (Chow et al., 2010; Watson et al., 2009) can pyrolyze during the analysis (Yang and Yu, 2002; Chow et al., 2004). ECR differs from ECT since the R signal is dominated by pyrolyzed OC (POC) on the filter surface, while the T signal is influenced by POC both on and within the filter (Chen et al., 2004; Chow et al., 2004). Unlike IMPROVE_A, which reports both ECR and ECT, other TOA protocols employ different temperature steps, often reporting only the ECT results (Birch and Cary, 1996; Cavalli et al., 2010; NIOSH, 1999; Schauer et al., 2003; Peterson and Richards, 2002).

The R and/or T measurements as part of TOA can infer the light absorption coefficient (b_{abs}), analogous to the principle of optical absorption monitors such as the aethalometer (Hansen et al., 1984), particle-soot absorption photometer (PSAP; Bond et al., 1999), and the multi-angle absorption photometer (MAAP; Petzold and Schönlinner, 2004). Both the aethalometer and PSAP apply T attenuation, while the MAAP incorporates both R and T attenuations in the calculation of b_{abs}. In any case, it is necessary to compensate for multiple-scattering and loading effects of the particle-filter matrix that cause deviations from the simple Beer's law (Chen et al., 2004; Arnott et al., 2005b; Virkkula et al., 2005). Black-carbon (BC) concentrations can be derived from b_{abs} by applying a mass- and wavelength-specific absorption efficiency (MAE_λ, typically in $m^2 g^{-1}$). Many collocated measurements show high correlations but different slopes in BC/EC comparisons (Ahmed et al., 2009; Quincey et al., 2009; Reisinger et al., 2008; Snyder and Schauer, 2007; Chow et al., 2009). This confirms the overlapping concept of EC and BC but also signifies the complex nature of carbonaceous material and uncertainties in such measurements (Andreae and Gelencséer, 2006; Moosmüller et al., 2009; Petzold et al., 2013; Lack et al., 2014).

While BC (or EC) absorbs light strongly across the solar spectrum (300–1000 nm), some organic compounds that evolve in the OC step can also absorb light, especially at shorter wavelengths (< 600 nm). These compounds have been termed "brown carbon" (BrC) and are associated with the smoldering phase of biomass burning and some end products of secondary aerosol formation (Andreae and Gelencséer, 2006; Clarke et al., 2007; Zhang et al., 2011). The spectral dependence of b_{abs} is often described by α:

$$\alpha(\lambda) = -\frac{d\ln(b_{abs}(\lambda))}{d\ln(\lambda)}, \tag{1}$$

where $\alpha(\lambda)$ is the absorption Ångström exponent. For BC (or EC) with graphitic-like structure and a constant refractive index, $\alpha = 1$ and b_{abs} is proportional to λ^{-1}. For BrC and mineral dust, α varies with λ and is mostly > 1 (Moosmüller et al., 2009; Chen et al., 2015), causing b_{abs} to increase more rapidly towards shorter wavelengths (blue and ultraviolet) than is the case for BC. The aerosol deposit thus appears to be brown – or sometimes yellow, red, or chartreuse – as the longer wavelengths of illuminating light are reflected and the shorter wavelengths are absorbed. The multi-wavelength aethalometer has revealed different spectral patterns that are indicative of BC, BrC, dust, and their mixtures (Sandradewi et al., 2008; Favez et al., 2009; Yang et al., 2009).

Given the large number of samples per year analyzed by TOA worldwide, the optical data acquired as part of the analysis could be used, in addition to OC and EC, for studies relevant to source apportionment, human health, visibility, and climate. Described and characterized here is a retrofit of a TOA carbon analyzer that expands the single-wavelength R and T monitoring to seven wavelengths for the IMPROVE_A analysis, hereafter designated thermal spectral analysis (TSA). Equivalence of the OC and EC fractions from TOA and TSA is demonstrated for several source and receptor samples, and the wavelength dependence of the OC–EC split is investigated. An approach to report spectral b_{abs} (as absorption optical depth) and α for decoupling the BC and non-BC components is also introduced. As an aid to readers, all abbreviations used in this paper are listed in the Supplement (Table S1).

2 Instrument design and calibration

The 633 nm He/neon (Ne) laser in the DRI Model 2001 carbon analyzer (Chow et al., 2011; Chen et al., 2012) is replaced with a package of seven diode lasers with wavelengths (λ) of 405, 450, 532, 635, 780, 808, and 980 nm. The use of diode lasers substantially reduces the cost of the optical module. It also provides stronger signals than using a light-emitting diode (LED) in other designs (e.g., Hadley et al., 2008), despite being limited by the wavelengths commercially available. While the 635 nm approximates the He/Ne laser, other wavelengths were selected to cover the visible and near-infrared regions. The lasers are alternately pulsed (2 consecutive pulses per laser, 14 pulses per cycle) and lock-in amplified at a frequency of 30 Hz, resulting in two cycles and $\sim$ four pulses for each wavelength every second. The bifurcated fiber optic for delivering the He / Ne laser to the reflectance light pipe is replaced with an eight-furcated optical-fiber bundle, one for each of the lasers and the last for transferring the light reflected from the filter punch (0.5 cm^2) to a photodiode (Fig. 1). Another light pipe on the opposite side of the filter directs the transmitted light toward a separate photodiode detector.

Photodiode signals are acquired with an NI6216 data acquisition system (National Instruments, Austin, TX) at a rate of up to 100 000 data points per second. The system integrates the product of photodiode and reference (30 Hz square wave) signals every second to suppress noise (e.g., from 60 Hz power supply and oven glow, random noise, and baseline drift). The resulting integrals are reported as the spectral laser reflectance and transmittance (LR_λ and LT_λ, respectively). Example thermograms with LR_λ and LT_λ are illustrated in Fig. S1 of the Supplement.

LR_λ and LT_λ are relative terms depending on not only the optical properties of the sample but also laser intensity, the geometry of the laser/detector setup, and the response of the photodiode. They are related to absolute fil-

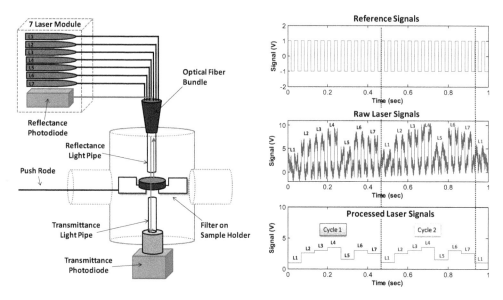

Figure 1. Configuration of optical sensing (left) modified from the DRI Model 2001 analyzer (Chen et al., 2012). The seven-laser module (i.e., L1-L7) represents seven diode lasers with wavelengths of 405, 450, 532, 635, 780, 808, and 980 nm. Reference (top right) and laser (middle right) signals are modulated at 30 Hz for a lock-in amplification of the reflectance or transmittance signals (bottom right). Only 30 Hz signals with the same phase as the reference are amplified at gains proportional to the reference signal voltage and number of data points integrated.

ter reflectance and transmittance (FR_λ and FT_λ, i.e., fraction of light reflected from or transmitted through the filter, respectively), using a set of transfer standards consisting of eight $PM_{2.5}$ quartz-fiber filters acquired using high-volume samplers from the Fresno Supersite (Watson et al., 2000; Chow et al., 2009), with EC loadings ranging from 0.9 to 15.8 $\mu g\, cm^{-2}$ (Fig. S2). The high-volume filters (406 cm^2) represent well-characterized urban aerosol, while providing sufficient sample for extensive testing. FR_λ and FT_λ of the transfer standards were predetermined using an integrating-sphere spectrometer (Lambda 35, Perkin Elmer, Massachusetts, USA; see, e.g., Zhong and Jang, 2011; Chen et al., 2015) traceable to NIST (National Institute of Standard and Technology) standards for wavelengths between 300 and 1000 nm.

Figure 2 compares LR_λ and LT_λ with FR_λ and FT_λ, respectively, for the transfer standards. Within the range of FR_λ (0.1–0.8) and FT_λ (0.0001–0.1), highly significant linear relationships ($r^2 > 0.9$) demonstrate the stability of the LR_λ and LT_λ measurements and the feasibility of converting them to FR_λ and FT_λ through calibration with a standardized spectrometer. Figure S1 illustrates the changes in FR_λ and FT_λ, retrieved from LR_λ and LT_λ, throughout a typical TSA. The uncertainty of the FR_λ and FT_λ retrievals may be evaluated by the coefficient of variance of the root mean square residual (CV-RMSR) in the FR_λ / LR_λ and FT_λ / LT_λ regression, respectively:

$$CV\text{-}RMSR = \frac{1}{Y_{obs}}\sqrt{\frac{\sum(Y_{cal} - Y_{obs})^2}{n-2}}, \quad (2)$$

where Y_{obs} is the FR_λ or FT_λ measured by the integrating-sphere spectrometer, Y_{cal} is the FR_λ or FT_λ calculated from LR_λ or LT_λ, respectively, and n is the number of samples. $FR_{633\,nm}$ and $FT_{633\,nm}$ based on the He / Ne laser of a typical Model 2001 analyzer exhibit a $\sim 3\%$ uncertainty (Fig. 2). For this retrofit, FR_λ uncertainties range from 3 to 11 % and FT_λ uncertainties range from 5 to 18 %, with the best precision shown at 450 and 808 nm. The divergence in uncertainty is attributed to the quality of the laser and the sensitivity of the photodiode detector for different wavelengths. Ongoing efforts to improve the signal-to-noise ratio of R and T measurements by using different lasers and averaging algorithms will be documented in subsequent papers.

3 Consistency of the OC–EC split

Several source and ambient $PM_{2.5}$ samples (Table 1) are used for testing the system. Samples from the Fresno Supersite (2 April 2003–28 December 2003) contain a mixture of carbonaceous materials from fresh engine exhaust, biomass burning, and cooking (Chen et al., 2007; Chow et al., 2007b; Gorin et al., 2006; Schauer and Cass, 2000). Reno ambient samples were acquired during the period of Rim Fire (17 August 2013–24 October 2013) and are dominated by an aged biomass burning plume containing mixed flaming and smoldering emissions. Source testing as part of the Lake Tahoe Prescribed Burning Study (Malamakal et al., 2013) and Gasoline/Diesel Split Study (Fujita et al., 2007) provided pure biomass burning and diesel exhaust samples, respec-

Table 1. Comparison of TC, OCR, OCT, ECR, and ECT between the 633 nm (TOA) and 635 nm (TSA) optical splits following the IMPROVE_A protocol.[a]

Sample type	Optical split (nm)[b]		Sample no. and avg.[c]			Corr.	Deming regression: $y = m \times x + b$		RD: $2(y-x)/(y+x)$	
	x	y	n	x	y	r^2	$m^c (\pm 1\sigma^d)$	$b^c (\pm 1\sigma^d)$	mean $\pm 1\sigma$	p^e
TC										
Fresno Supersite	635	633	10	20.30	20.05	1.00	0.97 ± 0.05	0.38 ± 0.75	-0.01 ± 0.04	0.56
Reno wildfire	635	633	14	30.69	30.21	1.00	0.91 ± 0.06	2.14 ± 1.18	0.04 ± 0.09	0.15
Prescribed burn	635	633	9	19.27	20.11	1.00	1.06 ± 0.09	-0.32 ± 1.12	0.01 ± 0.14	0.91
Diesel exhaust	635	633	11	8.10	7.78	0.97	0.99 ± 0.05	-0.26 ± 0.48	-0.06 ± 0.15	0.21
All	635	633	44	20.34	20.23	0.99	0.95 ± 0.07	0.89 ± 1.00	0.00 ± 0.12	0.85
OCR										
Fresno Supersite	635	633	10	16.02	15.91	0.99	0.97 ± 0.06	0.44 ± 0.76	0.00 ± 0.05	0.85
Reno wildfire	635	633	14	27.62	26.99	1.00	0.90 ± 0.07	2.09 ± 1.27	0.03 ± 0.09	0.30
Prescribed burn	635	633	9	17.22	18.25	1.00	1.07 ± 0.06	-0.12 ± 0.76	0.03 ± 0.13	0.65
Diesel exhaust	635	633	11	4.08	3.91	0.52	0.95 ± 0.35	0.04 ± 1.35	-0.04 ± 0.17	0.46
All	635	633	44	16.97	16.91	0.99	0.94 ± 0.08	0.90 ± 0.99	0.00 ± 0.12	0.68
OCT										
Fresno Supersite	635	633	10	17.80	17.82	1.00	1.01 ± 0.03	-0.17 ± 0.43	0.00 ± 0.04	0.85
Reno wildfire	635	633	14	29.07	28.64	1.00	0.92 ± 0.06	1.93 ± 1.19	0.03 ± 0.08	0.14
Prescribed burn	635	633	9	18.11	18.92	1.00	1.07 ± 0.05	-0.37 ± 0.72	0.01 ± 0.13	0.65
Diesel exhaust	635	633	11	4.42	4.14	0.46	1.01 ± 0.46	-0.33 ± 1.91	-0.07 ± 0.20	0.32
All	635	633	44	18.10	18.07	0.99	0.96 ± 0.07	0.74 ± 0.94	-0.01 ± 0.13	0.75
ECR										
Fresno Supersite	635	633	10	4.28	4.15	1.00	0.99 ± 0.06	-0.11 ± 0.17	-0.06 ± 0.08	0.06
Reno wildfire	635	633	14	3.07	3.22	0.99	1.04 ± 0.07	0.02 ± 0.15	0.12 ± 0.23	0.14
Prescribed burn	635	633	9	2.05	1.86	0.88	0.92 ± 0.18	-0.03 ± 0.34	-0.13 ± 0.25	0.25
Diesel exhaust	635	633	11	4.02	3.87	0.99	0.97 ± 0.07	-0.03 ± 0.23	-0.17 ± 0.30	0.12
All	635	633	44	3.37	3.31	0.99	1.00 ± 0.03	-0.05 ± 0.08	-0.04 ± 0.25	0.17
ECT										
Fresno Supersite	635	633	10	2.50	2.23	0.99	0.78 ± 0.12	0.29 ± 0.22	-0.07 ± 0.14	0.13
Reno wildfire	635	633	14	1.62	1.57	0.98	0.82 ± 0.02	0.24 ± 0.07	0.10 ± 0.30	0.33
Prescribed burn	635	633	9	1.16	1.18	0.85	0.87 ± 0.10	0.17 ± 0.13	0.07 ± 0.30	0.50
Diesel exhaust	635	633	11	3.68	3.64	0.99	0.92 ± 0.13	0.24 ± 0.30	-0.08 ± 0.33	0.70
All	635	633	44	2.24	2.16	0.98	0.90 ± 0.03	0.14 ± 0.06	0.01 ± 0.28	0.74

[a] TC: total carbon; OCR: organic carbon by reflectance; OCT: organic carbon by transmittance; ECR: elemental carbon by reflectance; and ECT: elemental carbon by transmittance following the IMPROVE_A thermal/optical carbon analysis protocol (Chow et al., 2007a). TOA: thermal/optical analyses; TSA: thermal/spectral analyses. [b] x is by retrofitted seven-wavelength carbon analyzer; y is by conventional single-wavelength (633 nm) DRI Model 2001 thermal/optical carbon analyzer. [c] Concentration in $\mu g\,cm^{-2}$; m is the slope; b is the intercept in $\mu g\,cm^{-2}$. [d] σ: standard deviation. [e] Student's t test p values.

tively. All these samples were analyzed by both TSA (using the retrofit) and TOA (using conventional Model 2001 analyzers) following the IMPROVE_A protocol.

Table 1 compares total carbon (TC), OC, and EC by reflectance (i.e., OCR and ECR) and transmittance (i.e., OCT and ECT) between TSA with the 635 nm and TOA with the normal 633 nm OC–EC split. As expected, TC is equivalent, with the averages agreeing within $\pm 5\%$ and regression slopes (m) ranging from 0.91 and 1.06 for each of the four sample types. The relative difference (RD), defined as the ratio of the difference divided by the average of two measurements (i.e., TSA and TOA) on the same sample, does not differ from 0 at the 5% significance level ($p > 0.05$). The

standard deviations of RD, a measure of random error, are higher for source (14–15%) than for ambient (4–9%) samples, indicative of greater deposit inhomogeneity for these samples, possibly due to variable sampling conditions over short sampling durations.

TSA and TOA also yield statistically equivalent OC and EC results, either by R or T (Table 1). Figure 3 visualizes the comparisons. With respect to the standard deviation of RD, OCR and OCT are similar to TC, while ECR and ECT are higher (up to 33%) due to a lower fraction of EC in TC. By category average, ECR and ECT account for 10–50 and 5–46% of TC, respectively. In general, ECR > ECT, as reported in previous studies (Khan et al., 2012; Han et al., 2013; Chow

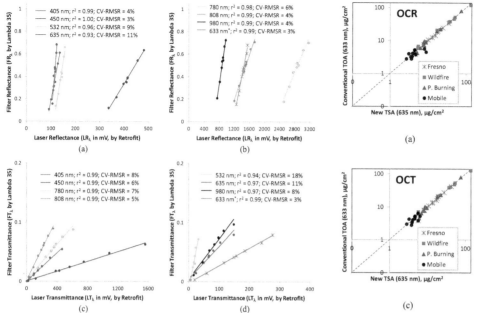

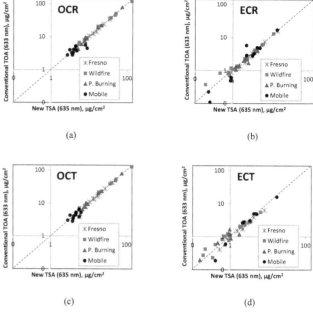

Figure 2. Calibration of spectral laser (**a, b**) reflectance and (**c, d**) transmittance measurements (LR_λ and LT_λ, respectively, in milli-volts (mV)) from the retrofitted seven-wavelength carbon analyzer at room temperature against absolute filter reflectance and transmittance (FR_λ and FT_λ, respectively) quantified by the Lambda integrating-sphere spectrometer, using eight Fresno ambient samples (5/6, 6/6, 6/19, 7/3, 9/29, 11/4, 11/13, and 12/28 of 2003) of various loadings as transfer standards. The 633 nm data are from a conventional carbon analyzer. r^2 and CV-RMSR evaluate the regression performance.

Figure 3. Comparison of organic and elemental carbon by reflectance optical correction (i.e., OCR and ECR, respectively) between the conventional thermal/optical analysis (TOA based on 633 nm optical split) and seven-wavelength thermal/spectral analysis (TSA based on 635 nm optical split) following the IMPROVE_A protocol. Note that x and y axes are on log scales.

et al., 2001; Schmid et al., 2001), consistent with organic vapors pyrolyzed within the filter leaving the sample after native EC and POC in the surface deposit have evolved (Chen et al., 2004). POC was least apparent for the diesel exhaust samples where optical adjustments were negligible.

The basic assumptions for optical adjustment include the following: (1) OC does not absorb light and (2) POC has the same apparent MAE as EC. Within-the-filter POC shows no apparent MAE by R (i.e., cannot detect it for any wavelengths) but high MAEs by T (higher than EC due to a multiple-scattering effect (Chow et al., 2004; Subramanian et al., 2006)). This leads to divergent ECR and ECT results. Since the multiple-scattering effect is larger for shorter wavelengths, ECT is expected to decrease with the wavelength at which the split is made. On the other hand, Chow et al. (1993) observed that the operational definition for EC by any TOA protocol might contain some light-absorbing OC. BrC, if present, would lower the baselines of R and T (prior to thermal analysis) for shorter wavelengths, resulting in earlier split points and larger EC reported than that from longer wavelengths.

Table 2 compares OC–EC splits for 635 nm with splits derived from shorter (450 nm) and longer (808 nm) wave-

lengths that have the lowest LR or LT uncertainties. ECRs based on the 635 nm split agree well with those based on the 808 nm split (i.e., slopes within the standard error from unity and p(RD) > 0.02) for all the sample types. This demonstrates that the 633 nm R split commonly used for optical adjustment since Huntzicker et al. (1982) does not significantly respond to BrC absorption compared with longer visible or infrared wavelengths. For ECT, the 880 nm split yielded higher values than the 635 nm split, especially for biomass-burning-dominated samples. This is consistent with ECT being influenced by POC within the filter and decreasing ECT with decreasing wavelength.

The 450 nm split appears to be sensitive to BrC content, as evidenced by larger ECR than those determined with 635 nm (except for diesel exhaust samples, where BrC contents are low). The difference is largest for the aged Reno wildfire samples, which may also contain some secondary organic aerosol formed during transport over the 200 km distance between Yosemite National Park and Reno, Nevada. As demonstrated in Fig. S1, shorter wavelengths produce earlier splits, resulting in higher ECR concentrations. The increase of ECT due to a 450 nm optical correction is not evident, which may be due to the opposing effects of BrC and POC at the shorter wavelength.

Table 2. Comparison of TSA-determined 635 nm ECR and ECT with 880 nm and 450 nm following the IMPROVE_A protocol.

Sample type	Optical split (nm)[a]		Sample no. and avg.[b]			Corr.	Deming regression: $y = m \times x + b$		RD: $2(y-x)/(y+x)$	
	x	y	n	x	y	r^2	$m^{\text{b}}(\pm 1\sigma^{\text{c}})$	$b^{\text{b}}(\pm 1\sigma^{\text{c}})$	mean$\pm 1\sigma$	p^{d}
ECR										
Fresno Supersite	635	450	10	4.28	4.74	1.00	1.23 ± 0.04	-0.55 ± 0.13	0.01 ± 0.13	0.82
Reno wildfire	635	450	14	3.07	4.29	0.95	1.90 ± 0.58	-1.55 ± 1.24	0.14 ± 0.25	0.06
Prescribed burn	635	450	9	2.05	2.14	0.98	1.41 ± 0.23	-0.74 ± 0.39	-0.03 ± 0.16	0.57
Diesel exhaust	635	450	11	4.02	3.93	1.00	1.00 ± 0.01	-0.10 ± 0.06	-0.08 ± 0.11	0.06
All	635	450	44	3.37	3.86	0.90	1.36 ± 0.28	-0.71 ± 0.66	0.02 ± 0.19	0.76
Fresno Supersite	635	808	10	4.28	4.12	1.00	0.95 ± 0.08	0.06 ± 0.22	-0.05 ± 0.05	0.03
Reno wildfire	635	808	14	3.07	3.22	1.00	1.04 ± 0.08	0.03 ± 0.17	0.15 ± 0.23	0.02
Prescribed burn	635	808	9	2.05	2.12	0.99	1.00 ± 0.09	0.06 ± 0.17	0.03 ± 0.09	0.30
Diesel exhaust	635	808	11	4.02	4.00	1.00	1.00 ± 0.00	-0.02 ± 0.02	-0.01 ± 0.02	1.00
All	635	808	44	3.37	3.40	1.00	0.99 ± 0.03	0.05 ± 0.07	0.04 ± 0.16	0.19
ECT										
Fresno Supersite	635	450	10	2.50	2.44	1.00	1.00 ± 0.02	-0.05 ± 0.07	-0.04 ± 0.08	0.16
Reno wildfire	635	450	14	1.62	1.35	0.99	0.85 ± 0.06	-0.02 ± 0.06	-0.20 ± 0.14	0.00
Prescribed burn	635	450	9	1.16	1.07	0.98	1.06 ± 0.25	-0.16 ± 0.22	-0.08 ± 0.24	0.16
Diesel exhaust	635	450	11	3.68	3.59	1.00	0.97 ± 0.01	0.01 ± 0.04	-0.04 ± 0.06	0.07
All	635	450	44	2.24	2.10	0.99	0.97 ± 0.02	-0.08 ± 0.03	-0.10 ± 0.16	0.00
Fresno Supersite	635	808	10	2.50	2.67	1.00	1.03 ± 0.04	0.10 ± 0.08	0.08 ± 0.04	0.00
Reno wildfire	635	808	14	1.62	2.00	0.99	1.24 ± 0.04	-0.01 ± 0.08	0.18 ± 0.18	0.00
Prescribed burn	635	808	9	1.16	1.27	0.99	1.10 ± 0.12	-0.01 ± 0.12	-0.09 ± 0.60	0.13
Diesel exhaust	635	808	11	3.68	3.53	1.00	0.99 ± 0.01	-0.11 ± 0.07	-0.24 ± 0.41	0.01
All	635	808	44	2.24	2.38	0.99	1.01 ± 0.07	0.11 ± 0.12	0.00 ± 0.38	0.02

[a] x and y are both by modified seven-wavelength carbon analyzer but with optical pyrolysis adjustment at different wavelengths. [b] Concentration in $\mu g\,cm^{-2}$; m is the slope; b is the intercept in $\mu g\,cm^{-2}$. [c] σ: standard deviation. [e] Student's t test p values.

4 Multi-wavelength absorption retrieval

Light absorption by particles on the filter is often estimated by transmittance attenuation (ATN):

$$\text{ATN}_\lambda = -\ln\left(\frac{\text{FT}_{\lambda,\text{i}}}{\text{FT}_{\lambda,\text{f}}}\right), \tag{3}$$

where i and f indicate FT_λ measured before and after thermal analysis, respectively. $\text{FT}_{\lambda,f}$ approximates a blank filter since all of the carbon has been removed. ATN_λ includes scattering and absorption within the substrate. On the other hand, the absorption optical depth ($\tau_{a,\lambda}$) measures only the light absorption. For diesel soot samples with negligible BrC and POC,

$$\tau_{a,\lambda} = \text{MAE}_{\lambda,\text{EC}} \times [\text{EC}], \tag{4}$$

where [EC] is the areal concentration of EC on filter in $\mu g\,cm^{-2}$. If there were no filter effects, ATN and τ_a would be identical for a given λ as described by Beer's law. With the knowledge of sampling volume (V) and filter area (A), ambient b_{abs} can be calculated from $\tau_a \times A/V$.

The relationship of [EC] and $\text{ATN}_{635\,\text{nm}}$ throughout the EC2 step (740 °C in a 98 % He / 2 % O_2 atmosphere) of the

IMPROVE_A analysis for a diesel exhaust sample is shown in Fig. 4. Since in this case carbon evolved during EC2 is exclusively EC, the temporal variation in [EC], i.e., d[EC] / dt, can be determined from carbon released and detected by the carbon analyzer. Arnott et al. (2005b) proposed a quadratic relationship between [EC] and ATN, derived from

$$\frac{\text{dATN}_\lambda}{\text{d}t} = \frac{M_\lambda \times \text{MAE}_{\lambda,\text{EC}}}{\sqrt{1 + \beta_\lambda \times [\text{EC}]}} \times \frac{\text{d}[\text{EC}]}{\text{d}t} \tag{5}$$

and thus

$$[\text{EC}] = \frac{\beta_\lambda}{4(M_\lambda \times \text{MAE}_{\lambda,\text{EC}})^2}\text{ATN}_\lambda{}^2$$
$$+ \frac{1}{M_\lambda \times \text{MAE}_{\lambda,\text{EC}}}\text{ATN}_\lambda, \tag{6}$$

where M_λ and β_λ account for the wavelength-specific multiple-scattering and loading effects, respectively. Eq. (6) fits the relationship in Fig. 4 well ($r^2 > 0.99$), thereby allowing M_λ and β_λ to be estimated. EC (quantified by IMPROVE_A ECR or ECT) of all the diesel exhaust samples exhibits a consistent dependence on the initial sample $\text{ATN}_{635\,\text{nm}}$ (prior to TSA) (Fig. 4). This supports the use of $\text{ATN}_{635\,\text{nm}}$ as a surrogate for EC and light absorption. Simi-

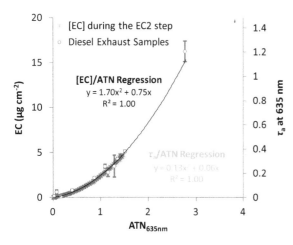

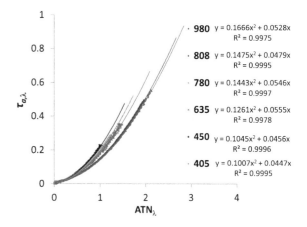

Figure 4. Relationship of EC concentration, [EC], versus $ATN_{635\,nm}$ from all 1 s measurements during the EC2 step (740 °C in a 98 % He / 2 % O_2 atmosphere) of IMPROVE_A analysis for a diesel exhaust sample (CIFQ074 from the Gas/Diesel Split Study (Fujita et al., 2007)). Also shown are paired [EC]–$ATN_{635\,nm}$ of 11 diesel exhaust samples, where EC is determined from IMPROVE_A and $ATN_{635\,nm}$ from initial filter transmittance. Circles and error bars indicate the average and spread, respectively, of EC by transmittance and reflectance (i.e., ECT and ECR, respectively). $\tau_{a,635\,nm}$ was further calculated from a MAE of $7.4\,m^2\,g^{-1} \times [EC]$, and the regression result between $\tau_{a,635\,nm}$ and $ATN_{635\,nm}$ is shown in green.

Figure 5. $\tau_{a,\lambda}$–ATN_λ relationships for 405–980 nm wavelengths, based on a diesel exhaust reference sample. Results for 532 nm are not shown owing to lack of sufficient detector signal-to-noise ratio at this wavelength.

lar relationships hold between [EC] and ATN for other wavelengths (Fig. S3). The largest scatter observed for 532 nm corresponds to the highest uncertainty in $FT_{532\,nm}$ retrievals; it results from the slow laser response time to modulation and the low sensitivity of the transmittance photodiode detector at this wavelength, which will be addressed in a subsequent design of the retrofit optics.

To relate $\tau_{a,635\,nm}$ to $ATN_{635\,nm}$, a MAE (EC) of $7.4\,m^2\,g^{-1}$ at 635 nm was used. The MAE was derived from concurrent b_{abs}, by a photoacoustic sensor at 1047 nm, and EC, by IMPROVE_A, measurements during the Gasoline/Diesel Split Study (Arnott et al., 2005a) and assuming an α of 1 in Eq. (1). The quadratic relationship holds for all of the available wavelengths, as seen in Fig. 5. Light absorption is enhanced by the filter since any $\tau_{a,\lambda} < 1$ corresponds to a larger ATN_λ, with more amplification towards shorter wavelengths. These $\tau_{a,\lambda}$–ATN_λ relationships would apply to any samples, regardless of the nature of light-absorbing material (e.g., EC, BC, BrC, or mineral dust). Particle penetration depth and, to a lesser degree, single-scattering albedo and the asymmetry g factor can influence the $\tau_{a,\lambda}$–ATN_λ dependence, though for a similar sampling configuration and filter material/thickness with typical loading, the perturbation is expected to be small (Chen et al., 2004; Arnott et al., 2005b). Retrieval algorithms employing both R and T such as those in Petzold and Schönlinner (2004) should be developed in the future to utilize all the information available.

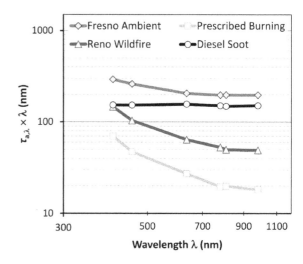

Figure 6. Product of absorption optical depth ($\tau_{a,\lambda}$) and wavelength (λ) as a function of λ by sample type. $\tau_{a,\lambda}$ shown represents averages over each of the sample types.

Using the relationships in Fig. 5, $\tau_{a,\lambda}$ was calculated for all samples from the initially measured ATN_λ values, with the average $\tau_{a,\lambda} \times \lambda$ by sample type compared in Fig. 6. The nearly constant $\tau_{a,\lambda} \times \lambda$ for diesel exhaust samples, i.e., $\tau_{a,\lambda} \propto \lambda^{-1}$, is consistent with the exclusive contribution of EC to light absorption. BC derived from the $\tau_{a,\lambda}$ would be equivalent to diesel EC. Averaged $\tau_{a,\lambda} \times \lambda$ increases by factors of 1.5, 3.0, and 3.9 from 980 to 405 nm for the Fresno ambient, Reno wildfire, and prescribed-burning samples, respectively. This reflects different levels of non-EC contribution.

Table 3. Average $\tau_{a,635\,nm}$ for four sample types and their respective BC and BrC fractions, BrC absorption Ångström exponent, and diesel-EC-equivalent BC (BC_d) concentration. IMPROVE_A ECR and ECT determined by the 635 nm optical adjustment are compared to BC_d in terms of average and p value of the relative difference (RD).

Sample type	$\tau_{a,635\,nm}$	$\tau_{a,635\,nm,BC}$	$\tau_{a,635\,nm,BrC}$ (% in τ_a)	α_{BC}^a	α_{BrC}	BC_d^b ($\mu g\,cm^{-2}$)	ECR ($\mu g\,cm^{-2}$)	ECT ($\mu g\,cm^{-2}$)	p value ($RD_{BC,ECR}$)	p value ($RD_{BC,ECT}$)
Fresno (ambient)	0.33	0.30	0.03 (10 %)	1	4.8 ± 1.5	4.0	4.3	2.5	0.02	0.87
Reno (wildfire)	0.10	0.07	0.02 (26 %)	1	4.8 ± 1.7	1.0	3.1	1.6	0.00	0.00
Prescribed burning	0.04	0.02	0.02 (46 %)	1	4.2 ± 1.8	0.3	2.1	1.2	0.00	0.00
Diesel exhausts	0.26	0.26	0.00 (1 %)	1	2.3 ± 0.1	3.5	4.0	3.7	0.05	0.06

[a] Pre-assumed values. [b] Calculated from $\tau_{a,635\,nm,BC}/(7.4\,m^2\,g^{-1})$.

5 Separation of BC and BrC contributions

A simplified two-component model consisting of BC and BrC, each with explicit absorption Ångström exponents (α_{BC} and α_{BrC}), is used to explain the spectral dependence of $\tau_{a,\lambda}$ in the samples:

$$\lambda_{a,\lambda} = q_{BC} \times \lambda^{-\alpha_{BC}} + q_{BrC} \times \lambda^{-\alpha_{BrC}}, \qquad (7)$$

where q_{BC} and q_{BrC} are fitting coefficients. This is analogous to the approach of Sandradewi et al. (2008), who considered the two components to be traffic and wood-burning particles and Hadley et al. (2008), who modeled two components of BC and char. Assuming an α_{BC} of 1 the same as diesel EC, then

$$\tau_{a,\lambda} \times \lambda = q_{BC} + q_{BrC} \times \lambda^{-(\alpha_{BrC}-1)}. \qquad (8)$$

Fitting coefficients in Eq. (8) were obtained for α_{BrC} values between 2 and 8 by least-square linear regression, and the α_{BrC} that led to the overall best fit in terms of r^2 was selected as the effective absorption Ångström exponent of BrC with which $\tau_{a,\lambda,BC}$ and $\tau_{a,\lambda,BrC}$ can be calculated from the first and second terms of Eq. (7). This fitting takes advantage of all six wavelengths. For each of the 44 samples, Fig. S4 shows that fitted $\tau_{a,\lambda}$ are within ± 5 % of the measured values for $\tau_{a,\lambda} > 0.01$. Examples of the $\tau_{a,\lambda}$ decomposition as a function of wavelength are shown in Fig. S5.

Table 3 summarizes the apportionment of $\tau_{a,635\,nm}$ into BC and BrC fractions along with average α_{BrC}, "diesel-EC-equivalent" BC (termed BC_d hereafter), ECR, and ECT by sample type. Consistent with Fig. 5, BrC contributions to $\tau_{a,635\,nm}$ are much higher in prescribed burning than in diesel exhaust samples (averaging 46 % versus 1 %) while somewhere in between (10–26 %) for Fresno and Reno wildfire samples. Effective α_{BrC} compares well among Fresno, Reno wildfire, and prescribed-burning samples (4.2–4.8) and is consistent with BrC of a similar nature from biomass burning (Bahadur et al., 2012; Kirchstetter and Thatcher, 2012). Even in the infrared region BrC accounts for 3, 6, and 24 % of $\tau_{a,980\,nm}$ for the Fresno, Reno, and prescribed-burning samples, respectively, on average. α_{BrC} in diesel exhaust, detectable in 5 of 11 samples, appears to be significantly lower (2.3 ± 0.1) than in other sample types.

Table 3 shows that BC_d as determined from $\tau_{a,635\,nm,BC}/MAE_{635\,nm}$ ($7.4\,m^2\,g^{-1}$) are lower than $ECR_{635\,nm}$. The differences are especially significant (i.e., p value of RD < 0.01) for Reno wildfire and Tahoe prescribed-burning samples with relatively high BrC contributions. The comparisons do not change with $ECR_{808\,nm}$ (with low BrC influence) replacing $ECR_{635\,nm}$. A continuum of light-absorbing carbon from biomass burning – ranging from BrC and char to soot – as suggested by Pöschl (2003) and Masiello (2004) may explain the phenomenon. As char and soot resulting from pyrolysis and high-temperature graphitization, respectively, are both quantified as EC by TOA (Han et al., 2009), they may have distinct optical properties. BC_d that was calibrated against diesel EC would represent just the soot fraction because there is little char material in diesel exhaust. $ECT_{635\,nm}$ is substantially lower than $ECR_{635\,nm}$ due to the aforementioned POC effect and is much closer to BC_d for the Reno and Tahoe biomass burning samples but not the Fresno and diesel exhaust samples.

6 Conclusions

Thermal/optical analysis that combines thermal separation and optical monitoring is potentially a powerful tool for analyzing carbonaceous aerosol on filters. Spatiotemporal variations and long-term trends in aerosol loading, chemical composition, sources, and effects have been inferred from OC and EC measurements (e.g., Chen et al., 2012; Hand et al., 2012; Malm et al., 1994; Murphy et al., 2011; Park et al., 2006). As many archived samples may be retrieved for reanalysis and $\sim 40\,000$ new samples are collected per year in the US long-term networks alone, an enhanced multi-wavelength thermal/optical analyzer would benefit the scientific community that uses the data.

The seven-wavelength (visible to near-infrared regions) TSA with both R and T sensors allows the determination of the OC–EC split at different wavelengths and light absorption measurements to be made with wavelength-specific loading corrections. In the selected ambient and source $PM_{2.5}$ samples, contributions of BC and BrC to light absorption were decoupled, assuming an absorption Ångström exponent of unity for BC and much higher values for BrC. Thus, BC concentrations optically equivalent to diesel exhaust EC, i.e., BC_d, can be calculated. BrC with an average absorption Ångström exponent of 4.2–4.8 is found to be enriched in samples influenced by biomass burning.

Despite the modifications in light source and detection technique, it is shown that the TSA measures $OC_{635\ nm}$ and $EC_{635\ nm}$ equivalent to $OC_{633\ nm}$ and $EC_{633\ nm}$ from conventional TOA following the same IMPROVE_A protocol with either R or T pyrolysis adjustment. $ECR_{635\ nm}$ is also consistent with those determined with longer wavelengths (e.g., 808 nm), though OC–EC splits with shorter wavelengths (e.g., 450 nm) increase ECR appreciably, showing the effect of BrC. For ECT, the BrC effect is somewhat canceled by an opposite POC effect. The optically derived BC_d underestimates $ECR_{635\ nm}$ or $ECR_{808\ nm}$ in biomass-burning-dominated samples with relatively high BrC content though the agreements are good for other samples. This discrepancy calls for further studies on the optical properties of EC, including soot and char, from biomass burning in contrast to those of diesel soot particles.

Acknowledgements. This work was supported, in part, by the US National Science Foundation (CHE 1214163), National Park Service IMPROVE Carbon Analysis Contract (C2350000894), and Chen's sabbatical leave at the University of Rostock. The authors thank Steve Kohl, Dana Trimble, and Gustavo Riggio at DRI for collection and carbon analysis of the samples and Megan Johnson for conducting the Lambda 35 spectral analysis. The conclusions are those of the authors and do not necessarily reflect the views of the sponsoring agencies.

Edited by: W. Maenhaut

References

Ahmed, T., Dutkiewicz, V. A., Shareef, A., Tuncel, G., Tuncel, S., and Husain, L.: Measurement of black carbon (BC) by an optical method and a thermal-optical method: Intercomparison for four sites, Atmos. Environ., 43, 6305–6311, 2009.

Andreae, M. O. and Gelencsér, A.: Black carbon or brown carbon? The nature of light-absorbing carbonaceous aerosols, Atmos. Chem. Phys., 6, 3131–3148, doi:10.5194/acp-6-3131-2006, 2006.

Arnott, W. P., Fujita, E. M., Walker, J., Campbell, D. E., Zielinska, B., Sagebiel, J. C., Moosmüller, H., Chow, J. C., and Lawson, D. R.: Photoacoustic Measurement of Black Carbon Emission Rates by Gasoline and Diesel Powered Vehicles and the Relationship with Carbon Analysis by Thermal Methods, in: Gasoline/Diesel

PM Split Study: Source and Ambient Sampling, Chemical Analysis, and Apportionment Phase, prepared by Desert Research Institute, Reno, NV, Prepared for DOE National Renewable Energy Laboratory, 2–39, 2005a.

Arnott, W. P., Hamasha, K., Moosmüller, H., Sheridan, P. J., and Ogren, J. A.: Towards aerosol light-absorption measurements with a 7-wavelength Aethalometer: Evaluation with a photoacoustic instrument and 3-wavelength nephelometer, Aerosol Sci. Technol., 39, 17–29, 2005b.

Bahadur, R., Praveen, P. S., Xu, Y. Y., and Ramanathan, V.: Solar absorption by elemental and brown carbon determined from spectral observations, Proc. Natl. Acad. Sci. USA, 109, 17366–17371, 2012.

Birch, M. E. and Cary, R. A.: Elemental carbon-based method for occupational monitoring of particulate diesel exhaust: Methodology and exposure issues, Analyst, 121, 1183–1190, 1996.

Bond, T. C., Anderson, T. L., and Campbell, D. E.: Calibration and intercomparison of filter-based measurements of visible light absorption by aerosols, Aerosol Sci. Technol., 30, 582–600, 1999.

Bougiatioti, A., Zarmpas, P., Koulouri, E., Antoniou, M., Theodosi, C., Kouvarakis, G., Saarikoski, S., Makela, T., Hillamo, R., and Mihalopoulos, N., Organic, elemental and water-soluble organic carbon in size segregated aerosols, in the marine boundary layer of the Eastern Mediterranean, Atmos. Environ., 64, 251–262, 2013.

Cao, J. J., Lee, S. C., Chow, J. C., Watson, J. G., Ho, K. F., Zhang, R. J., Jin, Z. D., Shen, Z. X., Chen, G. C., Kang, Y. M., Zou, S. C., Zhang, L. Z., Qi, S. H., Dai, M. H., Cheng, Y., and Hu, K.: Spatial and seasonal distributions of carbonaceous aerosols over China, J. Geophys Res.-Atmos., 112, 1–9, 2007.

Cavalli, F., Viana, M., Yttri, K. E., Genberg, J., and Putaud, J.-P.: Toward a standardised thermal-optical protocol for measuring atmospheric organic and elemental carbon: the EUSAAR protocol, Atmos. Meas. Tech., 3, 79–89, doi:10.5194/amt-3-79-2010, 2010.

Chen, L.-W. A., Chow, J. C., Watson, J. G., Moosmüller, H., and Arnott, W. P.: Modeling reflectance and transmittance of quartz-fiber filter samples containing elemental carbon particles: Implications for thermal/optical analysis, J. Aerosol Sci., 35, 765–780, 2004.

Chen, L.-W. A., Watson, J. G., Chow, J. C., and Magliano, K. L.: Quantifying $PM_{2.5}$ source contributions for the San Joaquin Valley with multivariate receptor models, Environ. Sci. Technol., 41, 2818–2826, 2007.

Chen, L.-W. A., Chow, J. C., Watson, J. G., and Schichtel, B. A.: Consistency of long-term elemental carbon trends from thermal and optical measurements in the IMPROVE network, Atmos. Meas. Tech., 5, 2329–2338, doi:10.5194/amt-5-2329-2012, 2012.

Chen, L.-W. A., Han, Y. M., Chow, J. C., Watson, J. G., and Cao, J. J.: Black carbon in dust and geological material: Optical analysis and implication of urban influence, J. Aerosol Sci., submitted, 2015.

Chow, J. C., Watson, J. G., Pritchett, L. C., Pierson, W. R., Frazier, C. A., and Purcell, R. G.: The DRI Thermal/Optical Reflectance carbon analysis system: Description, evaluation and applications in U.S. air quality studies, Atmos. Environ., 27A, 1185–1201, 1993.

Chow, J. C., Watson, J. G., Crow, D., Lowenthal, D. H., and Merrifield, T. M.: Comparison of IMPROVE and NIOSH carbon measurements, Aerosol Sci. Technol., 34, 23–34, 2001.

Chow, J. C., Watson, J. G., Chen, L.-W. A., Arnott, W. P., Moosmüller, H., and Fung, K. K.: Equivalence of elemental carbon by Thermal/Optical Reflectance and Transmittance with different temperature protocols, Environ. Sci. Technol., 38, 4414–4422, 2004.

Chow, J. C., Watson, J. G., Chen, L.-W. A., Chang, M.-C. O., Robinson, N. F., Trimble, D. L., and Kohl, S. D.: The IMPROVE_A temperature protocol for thermal/optical carbon analysis: Maintaining consistency with a long-term database, J. Air Waste Manage. Assoc., 57, 1014–1023, 2007a.

Chow, J. C., Watson, J. G., Lowenthal, D. H., Chen, L.-W. A., Zielinska, B., Mazzoleni, L. R., and Magliano, K. L.: Evaluation of organic markers for chemical mass balance source apportionment at the Fresno supersite, Atmos. Chem. Phys., 7, 1741–1754, 2007b, http://www.atmos-chem-phys.net/7/1741/2007/.

Chow, J. C., Watson, J. G., Doraiswamy, P., Chen, L.-W. A., Sodeman, D. A., Lowenthal, D. H., Park, K., Arnott, W. P., and Motallebi, N.: Aerosol light absorption, black carbon, and elemental carbon at the Fresno Supersite, California, Atmos. Res., 93, 874–887, 2009.

Chow, J. C., Watson, J. G., Chen, L.-W. A., Rice, J., and Frank, N. H.: Quantification of $PM_{2.5}$ organic carbon sampling artifacts in US networks, Atmos. Chem. Phys., 10, 5223–5239, doi:10.5194/acp-10-5223-2010, 2010.

Chow, J. C., Watson, J. G., Robles, J., Wang, X. L., Chen, L.-W. A., Trimble, D. L., Kohl, S. D., Tropp, R. J., and Fung, K. K.: Quality assurance and quality control for thermal/optical analysis of aerosol samples for organic and elemental carbon, Anal. Bioanal. Chem., 401, 3141–3152, 2011.

Clarke, A., McNaughton, C., Kapustin, V., Shinozuka, Y., Howell, S., Dibb, J., Zhou, J., Anderson, B., Brekhovskikh, V., Turner, H., and Pinkerton, M.: Biomass burning and pollution aerosol over North America: Organic components and their influence on spectral optical properties and humidification response, J. Geophys. Res.-Atmos., 112, D12S18, doi:10.1029/2006JD007777, 2007.

Dabek-Zlotorzynska, E., Dann, T. F., Martinelango, P. K., Celo, V., Brook, J. R., Mathieu, D., Ding, L. Y., and Austin, C. C.: Canadian National Air Pollution Surveillance (NAPS) $PM_{2.5}$ speciation program: Methodology and $PM_{2.5}$ chemical composition for the years 2003–2008, Atmos. Environ., 45, 673–686, 2011.

Favez, O., Cachier, H., Sciare, J., Sarda-Esteve, R., and Martinon, L.: Evidence for a significant contribution of wood burning aerosols to $PM_{2.5}$ during the winter season in Paris, France, Atmos. Environ., 43, 3640–3644, 2009.

Fujita, E. M., Zielinska, B., Campbell, D. E., Arnott, W. P., Sagebiel, J. C., Mazzoleni, L. R., Chow, J. C., Gabele, P. A., Crews, W., Snow, R., Clark, N. N., Wayne, W. S., and Lawson, D. R.: Variations in speciated emissions from spark-ignition and compression-ignition motor vehicles in California's south coast air basin, J. Air Waste Manage. Assoc., 57, 705–720, 2007.

Gorin, C. A., Collett Jr., J. L., and Herckes, P.: Wood smoke contribution to winter aerosol in Fresno, CA, J. Air Waste Manage. Assoc., 56, 1584–1590, 2006.

Hadley, O. L., Corrigan, C. E., and Kirchstetter, T. W.: Modified thermal-optical analysis using spectral absorption selectivity to distinguish black carbon from pyrolyzed organic carbon, Environ. Sci. Technol., 42, 8459–8464, 2008.

Han, Y. M., Lee, S. C., Cao, J. J., Ho, K. F., and An, Z. S.: Spatial distribution and seasonal variation of char-EC and soot-EC in the atmosphere over China, Atmos. Environ., 43, 6066–6073, 2009.

Han, Y. M., Chen, L.-W. A., Cao, J. J., Fung, K. K., Ho, K. F., Yan, B. Z., Zhan, C. L., Liu, S. X., Wei, C., and An, Z. H.: Thermal/optical methods for elemental carbon quantification in soils and urban dusts: Equivalence of different analysis protocols, Plos One, 8, e83462, doi:10.1371/journal.pone.0083462, 2013.

Hand, J. L., Schichtel, B. A., Pitchford, M. L., Malm, W. C., and Frank, N. H.: Seasonal composition of remote and urban fine particulate matter in the United States, J. Geophys Res.-Atmos., 117, D05209, doi:10.1029/2011JD017122, 2012.

Hansen, A. D. A., Rosen, H., and Novakov, T.: The aethalometer – An instrument for the real-time measurement of optical absorption by aerosol particles, Sci. Total Environ., 36, 191–196, 1984.

Huntzicker, J. J., Johnson, R. L., Shah, J. J., and Cary, R. A.: Analysis of organic and elemental carbon in ambient aerosols by a thermal-optical method, in: Particulate Carbon: Atmospheric Life Cycle, edited by: Wolff, G. T. and Klimisch, R. L., Plenum, New York, NY, USA, 79–88, 1982.

IMPROVE: Interagency Monitoring of Protected Visual Environments, prepared by National Park Service, Ft. Collins, CO, available at: http://vista.cira.colostate.edu/IMPROVE (last access: 16 January 2015), 2014.

Khan, B., Hays, M. D., Geron, C., and Jetter, J.: Differences in the OC/EC ratios that characterize ambient and source aerosols due to thermal-optical analysis, Aerosol Sci. Technol., 46, 127–137, 2012.

Kirchstetter, T. W. and Thatcher, T. L.: Contribution of organic carbon to wood smoke particulate matter absorption of solar radiation, Atmos. Chem. Phys., 12, 6067–6072, doi:10.5194/acp-12-6067-2012, 2012.

Lack, D. A., Moosmüller, H., McMeeking, G. R., Chakrabarty, R. K., and Baumgardner, D.: Characterizing elemental, equivalent black, and refractory black carbon aerosol particles: a review of techniques, their limitations and uncertainties, Anal. Bioanal. Chem., 406, 99–122, 2014.

Malamakal, T., Chen, L.-W. A., Wang, X. L., Green, M. C., Gronstal, S., Chow, J. C., and Watson, J. G.: Prescribed burn smoke impact in the Lake Tahoe Basin: model simulation and field verification, Int. J. Environ. Poll., 52, 225–243, 2013.

Malm, W. C., Sisler, J. F., Huffman, D., Eldred, R. A., and Cahill, T. A.: Spatial and seasonal trends in particle concentration and optical extinction in the United States, J. Geophys. Res., 99, 1347–1370, 1994.

Masiello, C. A.: New directions in black carbon organic geochemistry, Mar. Chem., 92, 201–213, 2004.

Moosmüller, H., Chakrabarty, R. K., and Arnott, W. P.: Aerosol light absorption and its measurement: A review, J. Quant. Spectrosc. Radiat. Transfer, 110, 844–878, 2009.

Murphy, D. M., Chow, J. C., Leibensperger, E. M., Malm, W. C., Pitchford, M., Schichtel, B. A., Watson, J. G., and White, W. H.: Decreases in elemental carbon and fine particle mass in the United States, Atmos. Chem. Phys., 11, 4679–4686, doi:10.5194/acp-11-4679-2011, 2011.

NIOSH: Method 5040 Issue 3 (Interim): Elemental carbon (diesel exhaust), in NIOSH Manual of Analytical Methods, 4th Edn.,

National Institute of Occupational Safety and Health, Cincinnati, OH, USA, 1999.

Park, R. J., Jacob, D. J., Kumar, N., and Yantosca, R. M.: Regional visibility statistics in the United States: Natural and transboundary pollution influences, and implications for the Regional Haze Rule, Atmos. Environ., 40, 5405–5423, 2006.

Peterson, M. R. and Richards, M. H.: Thermal-optical-transmittance analysis for organic, elemental, carbonate, total carbon, and OCX2 in PM2.5 by the EPA/NIOSH method, in: Proceedings, Symposium on Air Quality Measurement Methods and Technology-2002, edited by: Winegar, E. D. and Tropp, R. J., 83-1-83-19, Air & Waste Management Association, Pittsburgh, PA, USA, 2002.

Petzold, A. and Schönlinner, M.: Multi-angle absorption photometry – A new method for the measurement of aerosol light absorption and atmospheric black carbon, J. Aerosol Sci., 35, 421–441, 2004.

Petzold, A., Ogren, J. A., Fiebig, M., Laj, P., Li, S.-M., Baltensperger, U., Holzer-Popp, T., Kinne, S., Pappalardo, G., Sugimoto, N., Wehrli, C., Wiedensohler, A., and Zhang, X.-Y.: Recommendations for reporting "black carbon" measurements, Atmos. Chem. Phys., 13, 8365–8379, doi:10.5194/acp-13-8365-2013, 2013.

Pöschl, U.: Aerosol particle analysis: Challenges and progress, Anal. Bioanal. Chem., 375, 30–32, 2003.

Quincey, P. G., Butterfield, D., Green, D., Coyle, M., and Cape, C. N.: An evaluation of measurement methods for organic, elemental and black carbon in ambient air monitoring sites, Atmos. Environ., 43, 5085–5091, 2009.

Reisinger, P., Wonaschutz, A., Hitzenberger, R., Petzold, A., Bauer, H., Jankowski, N., Puxbaum, H., Chi, X., and Maenhaut, W.: Intercomparison of measurement techniques for black or elemental carbon under urban background conditions in wintertime: Influence of biomass combustion, Environ. Sci. Technol., 42, 884–889, 2008.

Sandradewi, J., Prevot, A. S. H., Szidat, S., Perron, N., Alfarra, M. R., Lanz, V. A., Weingartner, E., and Baltensperger, U.: Using aerosol light absorption measurements for the quantitative determination of wood burning and traffic emission contributions to particulate matter, Environ. Sci. Technol., 42, 3316–3323, 2008.

Schauer, J. J. and Cass, G. R.: Source apportionment of wintertime gas-phase and particle-phase air pollutants using organic compounds as tracers, Environ. Sci. Technol., 34, 1821–1832, 2000.

Schauer, J. J., Mader, B. T., Deminter, J. T., Heidemann, G., Bae, M. S., Seinfeld, J. H., Flagan, R. C., Cary, R. A., Smith, D., Huebert, B. J., Bertram, T., Howell, S., Kline, J. T., Quinn, P., Bates, T., Turpin, B. J., Lim, H. J., Yu, J. Z., Yang, H., and Keywood, M. D.: ACE-Asia intercomparison of a thermal-optical method for the determination of particle-phase organic and elemental carbon, Environ. Sci. Technol., 37, 993–1001, 2003.

Schmid, H. P., Laskus, L., Abraham, H. J., Baltensperger, U., Lavanchy, V. M. H., Bizjak, M., Burba, P., Cachier, H., Crow, D., Chow, J. C., Gnauk, T., Even, A., ten Brink, H. M., Giesen, K. P., Hitzenberger, R., Hueglin, C., Maenhaut, W., Pio, C. A., Puttock, J., Putaud, J. P., Toom-Sauntry, D., and Puxbaum, H.: Results of the "Carbon Conference" international aerosol carbon round robin test: Stage 1, Atmos. Environ., 35, 2111–2121, 2001.

Snyder, D. C. and Schauer, J. J.: An inter-comparison of two black carbon aerosol instruments and a semi-continuous elemental carbon instrument in the urban environment, Aerosol Sci. Technol., 41, 463–474, 2007.

Subramanian, R., Khlystov, A. Y., and Robinson, A. L.: Effect of peak inert-mode temperature on elemental carbon measured using thermal-optical analysis, Aerosol Sci. Technol., 40, 763–780, 2006.

Turpin, B. J., Cary, R. A., and Huntzicker, J. J.: An in-situ, time-resolved analyzer for aerosol organic and elemental carbon, Aerosol Sci. Technol., 12, 161–171, 1990.

U.S.EPA: Chemical speciation, prepared by U.S. Environmental Protection Agency, Research Triangle Park, NC, 2014.

Virkkula, A., Ahlquist, N. C., Covert, D. S., Arnott, W. P., Sheridan, P. J., Quinn, P. K., and Coffman, D. J.: Modification, calibration and a field test of an instrument for measuring light absorption by particles, Aerosol Sci. Technol., 39, 68–83, 2005.

Watson, J. G., Chow, J. C., Bowen, J. L., Lowenthal, D. H., Hering, S. V., Ouchida, P., and Oslund, W.: Air quality measurements from the Fresno Supersite, J. Air Waste Manage. Assoc., 50, 1321–1334, 2000.

Watson, J. G., Chow, J. C., and Chen, L.-W. A.: Summary of organic and elemental carbon/black carbon analysis methods and intercomparisons, Aerosol Air Quality Res., 5, 65–102, 2005.

Watson, J. G., Chow, J. C., Chen, L.-W. A., and Frank, N. H.: Methods to assess carbonaceous aerosol sampling artifacts for IMPROVE and other long-term networks, J. Air Waste Manage. Assoc., 59, 898–911, 2009.

Yang, H. and Yu, J. Z.: Uncertainties in charring correction in the analysis of elemental and organic carbon in atmospheric particles by thermal/optical methods, Environ. Sci. Technol., 36, 5199–5204, 2002.

Yang, M., Howell, S. G., Zhuang, J., and Huebert, B. J.: Attribution of aerosol light absorption to black carbon, brown carbon, and dust in China – interpretations of atmospheric measurements during EAST-AIRE, Atmos. Chem. Phys., 9, 2035–2050, doi:10.5194/acp-9-2035-2009, 2009.

Zhang, X. L., Lin, Y. H., Surratt, J. D., Zotter, P., Prevot, A. S. H., and Weber, R. J.: Light-absorbing soluble organic aerosol in Los Angeles and Atlanta: A contrast in secondary organic aerosol, Geophys. Res. Lett., 38, L21810, doi:10.1029/2011GL049385, 2011.

Zhang, X. Y., Wang, Y. Q., Niu, T., Zhang, X. C., Gong, S. L., Zhang, Y. M., and Sun, J. Y.: Atmospheric aerosol compositions in China: spatial/temporal variability, chemical signature, regional haze distribution and comparisons with global aerosols, Atmos. Chem. Phys., 12, 779–799, doi:10.5194/acp-12-779-2012, 2012.

Zhong, M. and Jang, M.: Light absorption coefficient measurement of SOA using a UV-Visible spectrometer connected with an integrating sphere, Atmos. Environ., 45, 4263–4271, 2011.

4

Determining the temporal variability in atmospheric temperature profiles measured using radiosondes and assessment of correction factors for different launch schedules

D. Butterfield and T. Gardiner

National Physical Laboratory, Hampton Road, Teddington, Middlesex, TW11 0LW, UK

Correspondence to: D. Butterfield (david.butterfield@npl.co.uk) and T. Gardiner (tom.gardiner@npl.co.uk)

Abstract. Radiosondes provide one of the primary sources of upper troposphere and stratosphere temperature data for numerical weather prediction, the assessment of long-term trends in atmospheric temperature, study of atmospheric processes and provide intercomparison data for other temperature sensors, e.g. satellites. When intercomparing different temperature profiles it is important to include the effect of temporal mismatch between the measurements. To help quantify this uncertainty the atmospheric temperature variation through the day needs to be assessed, so that a correction and uncertainty for time difference can be calculated. Temperature data from an intensive radiosonde campaign, at Manus Island in Papua New Guinea, were analysed to calculate the hourly rate of change in temperature at different altitudes and provide recommendations and correction factors for different launch schedules. Using these results, three additional longer term data sets were analysed (Lindenberg 1999 to 2008; Lindenberg 2009 to 2012; and Southern Great Plains 2006 to 2012) to assess the diurnal variability of temperature as a function of altitude, time of day and season of the year. This provides the appropriate estimation of temperature differences for given temporal separation and the uncertainty associated with them. A general observation was that 10 or more repeat measurements would be required to get a standard error of the mean of less than 0.1 K per hour of temporal mismatch.

1 Introduction

Radiosondes provide one of the primary sources of upper troposphere and stratosphere temperature data for numerical weather prediction, the assessment of long-term trends in atmospheric temperature, study of atmospheric processes and provide intercomparison data for other temperature sensors, e.g. satellites. For many of these applications, understanding the measurement uncertainty is crucial to effectively using the data and interpreting the relationship between different measurement sources. The Global Climate Observing System (GCOS) and the Reference Upper Air Network (GRUAN) have been established under the joint auspices of GCOS and relevant commissions of the World Meteorological Organization (WMO) as an international reference observing network, designed to meet climate requirements and to fill a large gap in the current global observing system (Thorne et al., 2013). Extensive work has been undertaken within GRUAN to establish the traceable measurement uncertainty associated with radiosonde measurements (Immler et al., 2010). However, when comparing profile results between different atmospheric sensors, the individual sensor measurement uncertainties only make up part of the overall comparison uncertainty. Allowance also has to be made for the coincidence uncertainty in time and space, and the smoothing uncertainty in the two profile measurements (von Clarmann, 2006). This paper addresses the coincidence uncertainty associated with using radiosonde results for intercomparisons with other temperature measurements.

Intercomparisons between temperature measurements made by radiosondes and satellites are well documented (Free and Seidel, 2005; Randel et al., 2009). The performance of radiosonde temperature sensors is reasonably well understood and these sensors are normally traceably calibrated on site before launch (Immler et al., 2010). Whereas satellite sensors are well characterised and calibrated before launch (Mo, 1996), there is no direct mechanism to validate this calibration post-launch or over the time history of the satellite's mission. Drift corrections can be performed (Zou and Wang, 2010) and agreements with other satellite measurement methods calculated (Zou et al., 2014), however these do not make a direct comparison with actual in-atmosphere temperature measurements. Regular intercomparisons between satellite and radiosonde measurements need to be performed to validate the ongoing temperature calibration of the satellite. Arranging a coincident satellite overpass of a radiosonde launch is difficult and in most cases impractical. Therefore the rate of change in atmospheric temperature needs to be assessed and an appropriate launch schedule determined to allow valid comparisons. Previous work (Sun et al., 2010) has found that the mean temperature difference (all altitudes) across 13 types of radiosonde and the Constellation Observing System for Meteorology, Ionosphere and Climate (COSMIC) Global Positioning System, Remote Occultation (GPS RO) satellite measurements for a global network to be 0.15 K. For Vaisala radiosondes, whose data was analysed in this paper, an increasing warm bias (from 0–0.4 K) with altitudes above 19 km (50 hPa) was found. The effect of the difference in radiosonde launch time and satellite overpass was also examined. The comparison standard deviation errors (all radiosonde types) for temperature were found to be 0.35 K per 3 h time difference.

The aim of this paper is to establish a methodology, from the limited data available from sites with a high launch frequency, to see if a data correction factor could be established at these sites to guide launch schedules. This represents the first step in developing a general tool for calculating temporal correction factors for any ground-based monitoring site. This work presents the results of a study of existing radiosonde data sets in order to estimate the uncertainty that would arise due to a temporal mismatch between a radiosonde profile and another source of temperature data. This is derived as a function of altitude, time of day and season of the year. This paper does not try to physically explain or quantify the reasons behind the correction factors derived at each site. In addition to providing an estimation of the coincidence uncertainty in time, it also gives guidance on the frequency of radiosonde launches required to capture diurnal variations.

2 Overview and data

To help quantify the difference between radiosonde and satellite measurements the diurnal atmospheric temperature variation needs to be assessed, so that a correction for time difference can be calculated. Radiosondes are routinely launched at 12-hourly intervals (00:00 and 12:00 UTC) from

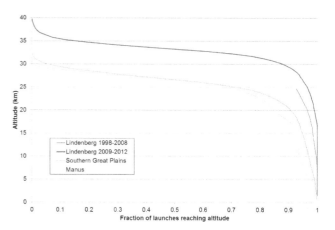

Figure 1. Fraction of radiosonde launches providing results as a function of altitude for each data set used. Blue line: Lindenberg 1999 to 2008; red line: Lindenberg 2009 to 2012; green line: Southern Great Plains; grey line: Manus Island.

many sites around the globe (Seidel and Free, 2006) with a very limited number of sites making more frequent measurements (WMO, 2013). To determine the frequency of launches needed to have an acceptable understanding of the atmosphere's temperature stability over short time periods (< 24 h), temperature measurements from radiosonde flights made by the upper-air sounding network for Dynamics of the Madden–Julian Oscillation (DYNAMO) at Manus Island, Papua New Guinea were analysed. During this campaign, Vaisala RS92-SGP radiosondes with GPS wind finding were launched every 3 h (00:00, 03:00, 06:00, 09:00, 12:00, 15:00, 18:00, 21:00 UTC) from 24 September 2011 to 31 March 2012. After conversion to local time, the hourly rate of change in temperature between launches was calculated for 500 m altitude bins from the surface to 24 km, for launches 3, 6 and 12 h apart. The mean hourly rates of change were inter-compared to assess the launch frequency required to acceptably characterise the diurnal change in temperature.

Following the analysis of launch frequency, long-term data from four radiosonde launches per day at Lindenberg (1999 to 2008 Vaisala RS90 radiosonde and 2009 to 2012 Vaisala RS92-SGP radiosonde) in Southern Germany and Southern Great Plains (Vaisala RS92 radiosonde 2006 to 2012) Oklahoma, USA were analysed to give hourly rates of change in temperature. Table 1 gives a summary of the radiosonde data sets.

The rate of change data was analysed to show differences in temperature stability between launches over a 24 h period and over the four seasons of the year, again up to an altitude of 24 km. Although some results were available up to 40 km, the number of samples fell off significantly with altitude – as shown in Fig. 1. The maximum altitude of 24 km was selected as a suitable upper limit as all data sets giving > 75 % data capture rates up to this altitude.

Table 1. Summary of radiosonde data sets used.

Launch site	Manus	Lindenberg	Lindenberg	Southern Great Plains
Latitude	$2°3'39.64''$ S	$52°12'36.0''$ N	$52°12'36.0''$ N	$36°36'18.0''$ N
Longitude	$147°25'31.43''$ E	$14°7'12.0''$ E	$14°7'12.0''$ E	$97°29'6.0''$ W
Start	24/09/2011	01/01/1999	01/01/2009	01/01/2006
End	31/03/2012	31/12/2008	31/12/2012	31/12/2012
Launches per day	8	4	4	4
Sonde	RS92-SGP	RS90	RS92-SGP	RS92-SGP
Total number of launches	1002	14466	4555	9754
Median near surface (0 to 500 m) temperature, °C	26.4	9.4		17.1
Minimum daily near surface (0 to 500 m) temperature, °C	15.9	−17.6		−14.3
Maximum daily near surface (0 to 500 m) temperature, °C	34.3	30.2		40.9

3 Results and discussion

3.1 Manus Island DYNAMO data set

Radiosonde temperature readings are amalgamated into altitude bins 500 m high, labelled as the centre of each bin, i.e. 0 to 500 m labelled as 250 m. The temperatures in each altitude bin are averaged to provide a mean temperature, T, for that specific altitude. The rate of change in temperature between single launches 3, 6 and 12 h apart, at each altitude, were calculated according to Eq. (1). The mean rate of change in temperature between each launch separation and altitude, $\frac{\mathrm{d}T}{\mathrm{d}t_n}$, were then calculated according to Eq. (2):

$$\frac{\mathrm{d}T}{\mathrm{d}t_n} = \frac{T_n - T_0}{t_n - t_0} \tag{1}$$

where $n = 3, 6$, or 12 h separation between launch time;

$$\overline{\frac{\mathrm{d}T}{\mathrm{d}t_n}} = \frac{\sum \frac{\mathrm{d}T}{\mathrm{d}t_n}}{i} \tag{2}$$

where $i =$ the number of launch pairs.

The mean rates of change in temperature ($\overline{\frac{\mathrm{d}T}{\mathrm{d}t_3}}$, $\overline{\frac{\mathrm{d}T}{\mathrm{d}t_6}}$ and $\overline{\frac{\mathrm{d}T}{\mathrm{d}t_{12}}}$) were used to define temperature change profiles over the day at different altitudes and are shown in Fig. 2. These profiles are for the complete 6-month data set and have not been split into seasons.

The times given in the figure show the mid-point in local time (LT) between the two launches used to calculate the temperature differences. Note that, for the 12 h separation results the launch times used are the 00:00 and 12:00 UTC radiosonde launches that are typically used by sites carrying out two launches per day. The error bars on the profiles come from the standard error of the mean. It can be seen in Fig. 2 that the profiles from launches 3 and 6 h apart follow similar profiles during the day, within the error bars (standard error of the mean), while the profiles from launches 12 h apart are unrepresentative and generally underestimate the actual diurnal variability. The profiles shown in Fig. 2 are a subset of all the altitudes evaluated. The complete set can be viewed online in the Supplement.

In order to quantify the difference between the different launch schedules it was assumed that 8 launches per day provided the best available measure of the changing state of the atmosphere. The mean hourly rates of change in temperature from these launches were therefore considered to be the base set. The difference in temperature change rates, $\Delta \frac{\mathrm{d}T}{\mathrm{d}t_n}$ (K h^{-1}) between the base set and a single launch, 2 launches a day and 4 launches a day were calculated according to Eq. (3). The results of which can be seen in Fig. 3:

$$\Delta \overline{\frac{\mathrm{d}T}{\mathrm{d}t_n}} = \frac{\sum \mathrm{ABS}\left(\overline{\frac{\mathrm{d}T}{\mathrm{d}t_n}} - \overline{\frac{\mathrm{d}T}{\mathrm{d}t_3}}\right)}{8} \tag{3}$$

where $n = 6$ or 12. For single launches, $\Delta \overline{\frac{\mathrm{d}T}{\mathrm{d}t_n}}$ was taken as the mean of $\overline{\frac{\mathrm{d}T}{\mathrm{d}t_3}}$.

It can be seen from Fig. 3 that there is a marked difference in the temperature change rate between 4 launches a day and 2 launches a day, and that there is little improvement in performing 2 launches a day over a single launch. The 4 launches per day data set is statistically different from the single launch data set at all altitudes except 3250 m, with a confidence level of 1σ (68 %). At the 2σ (95 %) level, three altitudes (9250, 12 250 and 15 250 m) are statistically different. It is therefore assumed in the later analyses that launches spaced 6 h apart provide a reasonable estimation of the hourly rate of change in temperature. Launches spaced 12 h apart do not suitably follow the short-term variations in temperature change over a 24 h period. Clearly this result only directly applies to the Manus data set, but it provides reasonable confidence in the use of 4 launches per day data for longer term analysis.

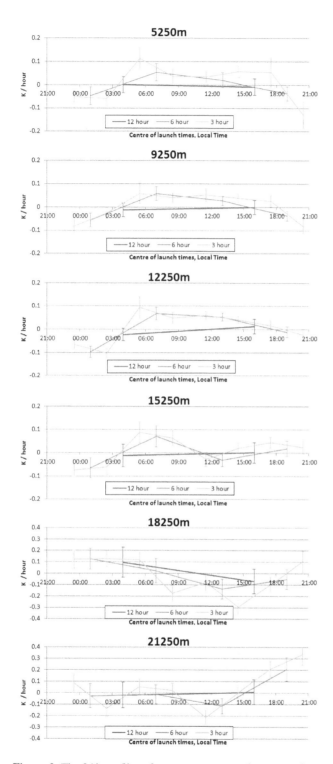

Figure 2. The 24 h profiles of mean temperature change rate from radiosonde launches at Manus Island during the DYNAMO campaign. Error bars are the standard error of the mean. Red line: 12 h separation; blue line: 6 h separation; green line: 3 h separation.

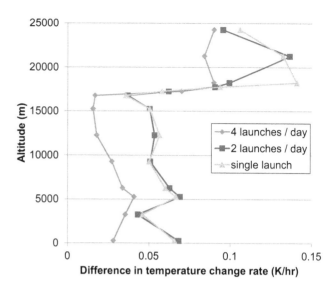

Figure 3. Difference in temperature change rate for a single launch (green triangles), 2 launches a day (red squares) and 4 launches a day (blue diamonds).

3.2   Lindenberg and Southern Great Plains data sets

Once a 6 h launch frequency was accepted to adequately represent the rate of change in temperature, three data sets were processed to calculate hourly rates of change, according to Eq. (2), between the four launches covering a 24 h period. Each data set was broken down into seasons and the calculations repeated to show if there were any changes in behaviour. Subsets of these results are shown in Figs. 4 and 5. Plots for all launches across all seasons can be viewed online in the Supplement. The error bars represent the standard error of the mean. Note that, as with the Manus data, the launches spaced 12 h apart (at 00:00 and 12:00 UTC) did not show the same degree of diurnal variability as the 6 h launch results.

It can be seen from Fig. 4 that all three data sets show similar behaviour for all launches during winter, except for the Lindenberg 1999–2008 data set, which shows some divergence in the stratosphere for the rate of change calculated from the 12:00 and 18:00 local time launches. Figure 5 shows the results for all four seasons of the rate of change calculated from the 12:00 and 18:00 local time launches. In addition to the winter divergence highlighted earlier, the Southern Great Plains (SGP) data set shows cooling in the stratosphere in spring, while the two Lindenberg data sets show heating. SGP shows significantly more heating in the troposphere and above 22 km in the summer. Autumn SGP results are also significantly different from Lindenberg in the lower troposphere, while the two Lindenberg data sets diverge in the stratosphere and are split by the SGP data set at this altitude. Summaries of near surface temperature (0–500 m) are given in Table 1 to give an indication of the local climate during the radiosonde launches. This difference in the stratosphere in the Lindenberg data may be due to the changes

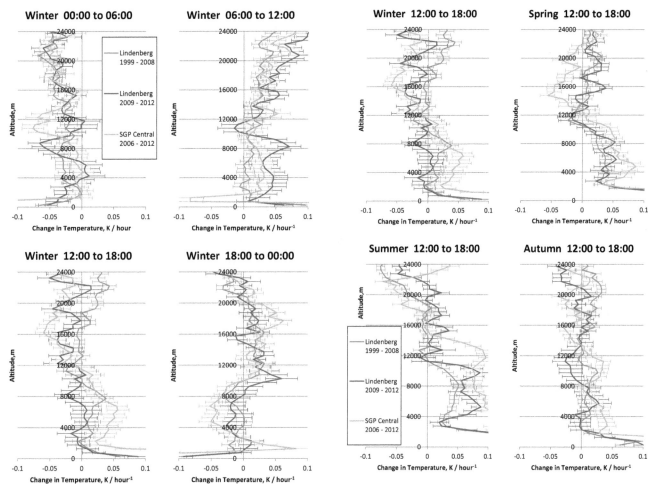

Figure 4. Hourly temperature change rate from 0 to 24 km, for the three data sets during winter. Blue line: Lindenberg 1999 to 2008; red line: Lindenberg 2009 to 2012; green line: Southern Great Plains.

Figure 5. Hourly temperature change rate from 0 to 24 km, for three data sets, calculated from launches at 12:00 and 18:00 local time for all four seasons. Blue line: Lindenberg 1999 to 2008; red line: Lindenberg 2009 to 2012; green line: Southern Great Plains.

in radiosonde type and analysis procedures between the two data sets. The influence of these changes and the effect of improved knowledge of the measurement uncertainty in the more recent data is a potential area for further investigation.

The error bars in Figs. 4 and 5 are expressed as the standard error of the mean result. If the standard deviation for a complete data is calculated and then the standard error calculated for differing numbers of repeat measurements, this gives an indication of the number of repeat measurements/radiosonde flights with corresponding satellite overpasses that would need to be made to bring the uncertainty in the temperature correction into acceptable bounds. Table 2 gives a summary of the mean temperature change rate between two launch times 6 h apart from a single data set (Lindenberg 1999 to 2008) along with the standard deviation of the measurements, the standard error of the mean for 10 and 100 repeated measurements for the four seasons of the

year. The results for the three data sets for all seasons can be viewed online in the Supplement.

Figure 6 summarises the results at 5 km altitude in spring for 13:00 and 19:00 local time, to give an indication of the reduction in the uncertainty with increased number of measurements for each data set. It can be seen that to obtain a standard error of the mean rate of change in temperature of $<=0.1$ K per hour, 10 or more repeat measurements are required. The standard errors of the means for 100 measurements in Table 2 are similar to those for the Manus Island results in Fig. 2 (0.038), which were typically made up of 90 launches per result. The number of launches per data point for the Lindenberg 1999 to 2008 data set is 889, for the Lindenberg 2009 to 2012 data set 227 and 572 for the Southern Great Plains data set.

The data in Fig. 6 also shows how these results could be used in practice. Taking the Lindenberg 1999 to 2008 results as an example, if a comparison was made between a single SGP radiosonde temperature measurement and another tem-

Table 2. Lindenberg 1999–2008. Mean rate of change in temperature between launches at 13:00 and 19:00 local time at different altitudes for each season, along with standard deviation of a single measurement and standard error with increased number of measurements.

Altitude 5 km	Spring	Summer	Autumn	Winter
Mean rate of change, $K\,h^{-1}$	0.036	0.040	0.010	0.013
SD (1 reading)	0.265	0.219	0.304	0.372
SE (10 readings)	0.084	0.069	0.096	0.118
SE (100 readings)	0.026	0.022	0.030	0.037
Altitude 10 km	Spring	Summer	Autumn	Winter
Mean rate of change, $K\,h^{-1}$	0.011	0.027	0.023	0.000
SD(1 reading)	0.305	0.280	0.337	0.368
SE (10 readings)	0.097	0.088	0.107	0.116
SE (100 readings)	0.031	0.028	0.034	0.037
Altitude 15 km	Spring	Summer	Autumn	Winter
Mean rate of change, $K\,h^{-1}$	0.006	−0.005	0.004	−0.003
SD (1 reading)	0.182	0.191	0.215	0.235
SE (10 readings)	0.058	0.060	0.068	0.074
SE (100 readings)	0.018	0.019	0.021	0.023
Altitude 20 km	Spring	Summer	Autumn	Winter
Mean rate of change, $K\,h^{-1}$	0.031	−0.033	0.032	0.024
SD (1 reading)	0.199	0.175	0.202	0.270
SE (10 readings)	0.063	0.055	0.064	0.085
SE (100 readings)	0.020	0.017	0.020	0.027

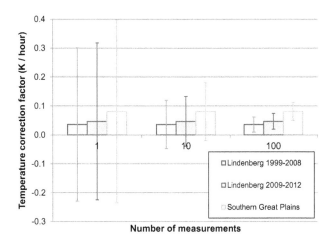

Figure 6. Reduction in uncertainty in hourly temperature change rate due to repeat radiosonde flights – for measurements between 13:00 and 19:00 LT, at 5 km altitude in spring. Columns show the mean temperature change rate and error bars should the uncertainty associated with different numbers of samples. Blue: Lindenberg 1999 to 2008; red: Lindenberg 2009 to 2012; green: Southern Great Plains.

perature measurement (between 13:00 and 19:00 local time, at 5 km, in spring) then for each hour difference between the measurements a correction of 0.036 K should be applied to the radiosonde result and an additional random uncertainty of

0.265 K should be included in the comparison. This correction should be subtracted from the radiosonde measurement to adjust for the temporal mismatch. If this was repeated 10 times the correction factor would remain the same, but the additional random uncertainty would reduce to 0.084 K. The supplementary data gives a summary of results for all three data sets over the separate 6 h launch separations. The results enable such an evaluation to be made for any altitude, time of day and season.

4 Conclusions and further work

Four radiosonde data sets have been analysed to assess the temporal variability of the temperature profile as a function of altitude, time of day and season of the year. This provides information on the temporal mismatch uncertainty that would result from comparing atmospheric temperature measurements at different times. The results from the intensive Manus campaign with 8 launches per day show that 2 radiosonde launches per day (at 00:00 and 12:00 UTC) do not capture the diurnal variability and would tend to underestimate both the adjustment and uncertainty that would result from a temporal mismatch, but that 4 radiosonde launches per day provides a reasonable estimate of the diurnal variability.

Analysis of longer term data sets with four launches per day provide appropriate estimation of temperature differences for a given temperature separation and the uncertainties associated with them. The uncertainties show similar behaviour for all data sets and indicate that, in general, 10 or more repeat measurements would be required to get a standard uncertainty of less than 0.1 K per hour of temporal mismatch.

Having established that the method presented in this paper is a viable one for estimating temporal variability, it should be recognised that these results only directly apply to the radiosonde launch sites from which the data sets have been obtained. In order to generate appropriate correction factors for other sites the method will require further development, using additional data sources or model results for each site.

Given the conclusion that at least four launches per day are needed to capture the diurnal variability and the very limited number of launch sites from which such long term data is available, then a modification to this analysis would be needed to give it wider global applicability. Two methods to consider are combining twice daily radiosonde results with higher temporal resolution data from another measurement technique or using high resolution meteorological models to fill in the gaps between the radiosonde launches. Both options will be the subject of further work.

Acknowledgements. Manus

One of the upper air data sets developed for the Dynamics of the Madden-Julian Oscillation (DYNAMO) 2011 to 2012 project. This data set includes 1411 high vertical resolution (2 s) soundings from the Atmospheric Radiation Measurement (ARM) C1 Momote. These data were provided by ARM and had preliminary quality control by NCAR/EOL. This L3 version of the data set has a correction by CSU. This station used Vaisala RS92-SGP radiosondes with GPS wind finding during the DYNAMO field campaign.

Data provided by NCAR/EOL under sponsorship of the National Science Foundation. http://data.eol.ucar.edu/. Dataset name: Manus ARM AMF Radiosonde L3 Data (ESC Format) [NCAR/EOL], http://data.eol.ucar.edu/codiac/dss/id=347.009.

Southern Great Plains

Data were obtained from the Atmospheric Radiation Measurement (ARM) Program sponsored by the US Department of Energy, Office of Science, Office of Biological and Environmental Research, Climate and Environmental Sciences Division.

Lindenberg

The 1999 to 2009 data is based on radiosonde measurements using Vaisala RS90 instruments.

The 2009 to 2012 data is a GRUAN data product (RS92-GDP V2) based on radiosonde measurements using Vaisala RS92 instruments. All GRUAN data products are based on measurements and processing that adheres to the GRUAN principles (Immler, 2010). The raw data are read from the original Digi-Cora III data base files and are corrected for known systematic biases. The uncertainty of the temperature, the humidity and the wind is calculated from estimates of the calibration uncertainty, the uncertainty of the bias correction and the statistical noise.

Data provided by German Meteorological Service (DWD).

Edited by: B. Kahn

References

Free, M. and Seidel, D.: Causes of differing temperature trends in radiosonde upper air data sets, J. Geophys. Res., 110, D07101, doi:10.1029/2004JD005481, 2005.

Immler, F. J., Dykema, J., Gardiner, T., Whiteman, D. N., Thorne, P. W., and Vömel, H.: Reference Quality Upper-Air Measurements: guidance for developing GRUAN data products, Atmos. Meas. Tech., 3, 1217–1231, doi:10.5194/amt-3-1217-2010, 2010.

Mo, T.: Prelaunch calibration of the Advanced Microwave Sounding Unit-A for NOAA-K, IEEE T. Microw. Theory, 44, 1460–1469, doi:10.1109/22.536029, 1996.

Randel, W. J., Shine, K. P., Austin, J., Barnett, J., Claud, C., Gillett, N. P., Keckhut, P., Langematz, U., Lin, R., Long, C., Mears, C., Miller, A., Nash, J., Seidel, D. J., Thompson, D. W. J., Wu, F., and Yoden, S.: An update of observed stratospheric temperature trends, J. Geophys. Res., 114, D02107, doi:10.1029/2008JD010421, 2009.

Seidel, D. J. and Free, M.: Measurement requirements for climate monitoring of upper air temperature derived from reanalysis data, J. Climate, 19, 854–871, 2006.

Sun, B., Reale, A., Seidel, D. J., and Hunt, D. C.: Comparing radiosonde and COSMIC atmospheric profile data to quantify differences among radiosonde types and the imperfect collocation on comparison statistics, J. Geophys. Res., 115, D23104, doi:10.1029/2010JD014457, 2010.

Thorne, P., Voemel, H., Bodeker, G., Sommer, M., Apituley, A., Berger, F., Bojinski, S., Braathen, G., Calpini, B., Demoz, B., Diamond, H. J., Dykema, J., Fasso, A., Fujiwara, M., Gardiner, T., Hurst, D., Leblanc, T., Madonna, F., Merlone, A., Mikalsen, A., Miller, C. D., Reale, T., Rannat, K., Richter, C., Seidel, D. J., Shiotani, M., Sisterson, D., Tan, D. G. H., Vose, R. S., Voyles, J., Wang, J., Whiteman, D. H., and Williams, S.: GCOS reference upper air network (GRUAN): steps towards assuring future climate records, AIP Conf. Proc., 1552, 1042, doi:10.1063/1.4821421, 2013.

von Clarmann, T.: Validation of remotely sensed profiles of atmospheric state variables: strategies and terminology, Atmos. Chem. Phys., 6, 4311–4320, doi:10.5194/acp-6-4311-2006, 2006.

WMO: The GCOS Reference Upper-Air Network (GRUAN) Guide, WIGOS Technical Report No. 2013-03, GCOS-171, 2013.

Zou, C. Z. and Wang, W.: Stability of the MSU derived atmospheric trend, J. Atmos. Ocean. Tech., 27, 1960–1971, doi:10.1175/2009JTECHA1333.1, 2010.

Zou, X., Lin, L., and Weng, F.: Absolute calibration of ATMS upper level temperature sounding channels using GPS RO observations, IEEE T. Geosci. Remote, 52, 1397–1406, doi:10.1109/TGRS.2013.2250981, 2014.

5

Impacts of cloud heterogeneities on cirrus optical properties retrieved from space-based thermal infrared radiometry

T. Fauchez[1], P. Dubuisson[1], C. Cornet[1], F. Szczap[2], A. Garnier[3,4], J. Pelon[5], and K. Meyer[6,7]

[1]Laboratoire d'Optique Atmosphérique, Université Lille 1, Villeneuve d'Ascq, France
[2]Laboratoire de Météorologie Physique, Université Blaise Pascal, Clermont Ferrand, France
[3]Science Systems and Applications, Inc., Hampton, Virginia, USA
[4]NASA Langley Research Center, Hampton, Virginia, USA
[5]Laboratoire Atmosphères, Milieux, Observations Spatiales, UPMC-UVSQ-CNRS, Paris, France
[6]Goddard Earth Sciences Technology and Research (GESTAR), Universities Space Research Association, Columbia, Maryland, USA
[7]NASA Goddard Space Flight Center, Greenbelt, Maryland, USA

Correspondence to: C. Cornet (celine.cornet@univ-lille1.fr)

Abstract. This paper presents a study, based on simulations, of the impact of cirrus cloud heterogeneities on the retrieval of cloud parameters (optical thickness and effective diameter) for the Imaging Infrared Radiometer (IIR) on board CALIPSO. Cirrus clouds are generated by the stochastic model 3DCLOUD for two different cloud fields and for several averaged cloud parameters. One cloud field is obtained from a cirrus observed on 25 May 2007 during the airborne campaign CIRCLE-2 and the other is a cirrus uncinus. The radiative transfer is simulated with the 3DMCPOL code. To assess the errors due to cloud heterogeneities, two related retrieval algorithms are used: (i) the split-window technique to retrieve the ice crystal effective diameter and (ii) an algorithm similar to the IIR operational algorithm to retrieve the effective emissivity and the effective optical thickness. Differences between input parameters and retrieved parameters are compared as a function of different cloud properties such as the mean optical thickness, the heterogeneity parameter and the effective diameter. The optical thickness heterogeneity for each $1\,\text{km} \times 1\,\text{km}$ observation pixel is represented by the optical thickness standard deviation computed using $100\,\text{m} \times 100\,\text{m}$ subpixels. We show that optical thickness heterogeneity may have a strong impact on the retrieved parameters, mainly due to the plane-parallel approximation (PPA assumption). In particular, for cirrus clouds with ice crystal diameter of approximately $10\,\mu\text{m}$, the averaged error on the retrieved effective diameter and optical thickness is about $2.5\,\mu\text{m}$ ($\sim 25\,\%$) and -0.20 ($\sim 12\,\%$), respectively. Then, these biases decrease with increasing effective size due to a decrease of the cloud absorption and, thus, the PPA bias. Cloud horizontal heterogeneity effects are greater than other possible sources of retrieval errors such as those due to cloud vertical heterogeneity impact, surface temperature or atmospheric temperature profile uncertainty and IIR retrieval uncertainty. Cloud horizontal heterogeneity effects are larger than the IIR retrieval uncertainty if the standard deviation of the optical thickness, inside the observation pixel, is greater than 1.

1 Introduction

In the context of global climate change, the representation and role of clouds are still uncertain. For example, ice clouds play an important role in the climate and on the Earth's radiation budget (Liou, 1986). Cirrus clouds lead mainly to a positive radiative forcing due to their high temperature contrast with respect to the surface. However, the cirrus radiative forcing could depend on the cirrus optical thickness, altitude and ice crystal effective size (Katagiri et al., 2013). Consequently, to improve our knowledge, it is essential to assess the feedback and climate effects of these clouds (Stephens,

1980). Global satellite observations are well suited to monitoring and investigating cloud evolution and characteristicsm because passive top-of-atmosphere (TOA) radiometric measurements allow for retrievals of cloud properties such as optical thickness and ice crystal effective diameter. In this work, we focus on infrared measurements obtained by the Imaging Infrared Radiometer (IIR; Garnier et al., 2012, 2013) onboard the Cloud-Aerosol Lidar and Infrared Pathfinder Satellite Observations (CALIPSO).

Because of operational constraints (lack of information regarding the 3-D structure of the atmosphere, time constraints, etc.), satellite-based cloud retrieval algorithms assume that clouds are homogeneous and infinite between two planes. This assumption of 1-D radiative transfer is called the homogeneous independent pixel approximation (Cahalan et al., 1994) or independent column approximation (Stephens et al., 1991). However, in a real atmosphere, clouds have 3-D structures, i.e, horizontal and vertical heterogeneities and the simplified 1-D atmosphere assumption may lead to biased cloud property retrievals (Fauchez et al., 2014). Many studies have been conducted to determine the impact of cloud heterogeneities on cloud products derived from solar spectral measurements. These studies primarily focused on warm clouds such as stratocumulus (Varnai and Marshak, 2001; Zinner and Mayer, 2006; Kato and Marshak, 2009, etc.) and showed that the sign and amplitude of retrieval errors depend on numerous factors, such as the spatial resolution, wavelength, geometry of observation and cloud morphology. Concerning cirrus clouds, Fauchez et al. (2014) showed that cirrus cloud heterogeneities lead to non-negligible effects on brightness temperatures (BT) and that these effects mainly depend on the standard deviation of the optical thickness inside the observation pixel. The retrieval of cloud properties using radiances or BT may thus be impacted by the heterogeneity effects. In this work, we extend the study of Fauchez et al. (2014) to investigate the impacts of cirrus heterogeneities on cloud optical property (optical thickness and ice crystal effective size) retrievals using simulations of radiometric measurements of IIR in three typical spectral bands, namely 8.65, 10.60 and 12.05 μm.

In the thermal infrared atmospheric window (8–13 μm), cloud optical properties (optical thickness and ice crystal effective size) are retrieved using the split-window technique (SWT) (Inoue, 1985; Parol et al., 1991; Dubuisson et al., 2008). This method is generally limited to thin cirrus clouds (optical thickness less than approximately 3 at 12 μm) and small crystals (effective diameters smaller than approximately 40 μm). In the visible and near-infrared spectra, cloud optical properties are commonly retrieved using the Nakajima and King method (Nakajima and King, 1990) that combines measurements in visible and near-infrared channels for optically thicker cirrus clouds and larger ice crystals. Cooper et al. (2007) combined these two methods for MODIS measurements to treat thin and thick cirrus simultaneously.

The paper is organized as follows. In Sect. 2, we present a short description of the modeling tools used in this study: (i) the cloud generator 3DCLOUD (Szczap et al., 2014), (ii) the radiative transfer code 3DMCPOL (Cornet et al., 2010; Fauchez et al., 2014) and (iii) two related retrieval algorithms. In Sect. 3, we present possible retrieval errors due to the 1-D approximation. In Sect. 4, we compare heterogeneity effects with other possible error sources considered in this paper such as those due to cloud vertical heterogeneity, surface temperature or atmospheric temperature profile uncertainty, as well as the IIR retrieval uncertainty. Conclusions and perspectives are given in Sect. 5.

2 Numerical models

2.1 3-D ice water content generation

The stochastic model 3DCLOUD (Cornet et al., 2010; Szczap et al., 2014) is employed to generate realistic 3-D cirrus clouds. This model uses a simplified dynamical and thermodynamical approach to generate heterogeneous 3-D clouds as well as a Fourier transform framework to constrain scale invariant properties (Hogan and Kew, 2005; Szczap et al., 2014). Two different cirrus fields were simulated (Fig. 1) in a mid-latitude summer (MLS) atmosphere.

The first cirrus field has been modeled from meteorological profiles presented by Starr and Cox (1985) coupled with a wind profile to form virgas. The cloud layer is defined by the mean optical thickness τ_c, the standard deviation of the optical thickness on the entire field σ_{τ_c}, the cirrus heterogeneity parameter $\rho_\tau = \sigma_{\tau_c}/\tau_c$ (Szczap et al., 2000) and the ice crystal effective diameter D_{eff} for an aggregate crystal shape (Yang et al., 2005). D_{eff} is defined as

$$D_{\mathrm{eff}} = \frac{3}{2} \frac{\int V(L)n(L)\mathrm{d}L}{\int A(L)n(L)\mathrm{d}L}, \qquad (1)$$

where L is the maximum crystal size, $V(L)$ is the volume of the crystal, $A(L)$ is the projected area and $n(l)$ is the size distribution (Yang et al., 2000).

Eight cirrus clouds are generated (Table 1) by varying the above parameters to cover the characteristics of typical cirrus clouds (Sassen and Cho, 1992; Szczap et al., 2000; Carlin et al., 2002; Lynch et al., 2002). Note that the effective diameter of cirrus cases 3 to 5 ($D_{\mathrm{eff}} = 9.95\,\mu$m) is probably too small for cirrus with a mean optical thickness of 1.80 because aggregations processes tend to increase the effective size (Fig. 12 of Garnier et al., 2013). Cloud heterogeneity effects are probably slightly overestimated due to the too-small crystal effective size (heterogeneity effects are larger for small effective sizes) with respect to the mean cirrus optical thickness. Nevertheless, cirrus cases 3 to 5 are useful for understanding how heterogeneity effects increase with the optical thickness heterogeneity parameter (ρ_τ in Table 1), which increases from 0.7 to 1.1 and 1.5 with other

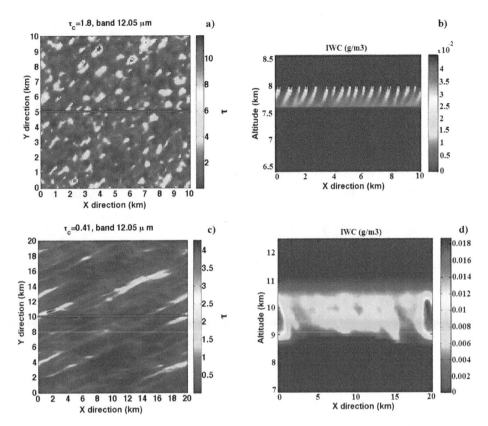

Figure 1. Top figures: Cirrus generated from realistic meteorological conditions (Starr and Cox, 1986; Hogan and Kew, 2005) with (**a**) the 10 km × 10 km optical thickness field simulated at 12.05 μm with a horizontal spatial resolution of 100 m and (**b**) the x–z view through the red line of (**a**) of the cirrus IWC with a vertical spatial resolution of 58 m. Bottom figures: CII cirrus simulation based on optical and microphysical properties of the cirrus observed during the CIRCLE-2 campaign on 25 May 2007: (**c**) the 20 km × 20 km optical thickness field at 12.05 μm, with a horizontal spatial resolution of 100 m and with a mean optical thickness $\tau_c = 0.41$ observed by IIR at 12.05 μm, and (**d**) the x–z view through the red line of (**c**) the cirrus IWC with a vertical resolution of 58 m.

cloud properties held constants. Two cirrus cloud cases are presented in Fig. 1. The first cloud structure is presented in Fig. 1a and b. Figure 1a presents the 10 km × 10 km optical thickness field at 12.05 μm with a spatial resolution of 100 m, and Fig. 1b presents the x–z view of the ice water content (IWC) of cirrus case 3.

Figure 1c and d show cirrus generated from measurements obtained on 25 May 2007 during the CIRCLE-2 airborne campaign (Mioche et al., 2010). In situ measurements provided by the aircraft, as well as IIR radiometric measurements (mean optical thickness and mean heterogeneity parameter), are used as input for 3DCLOUD. In addition, meteorological data from the European Center for Medium-Range Weather Forecasts are used to constrain the meteorological profiles (wind speed and orientation, temperature, humidity, etc.). The scale invariant properties of every cirrus case presented in Table 1 are controlled by a constant spectral slope ($-5/3$) for all scales and altitude levels. This agrees with the spectral slope of the backscattering coefficient measured at 532 nm at different altitudes by the Cloud-Aerosol Lidar with Orthogonal Polarization (CALIOP) on-board CALIPSO and

the extinction coefficient measured by the polar nephelometer at the aircraft altitude (Fauchez et al., 2014).

2.2 Optical property parametrization

Cirrus optical properties are difficult to characterize because of the diversity of crystal sizes, shapes and orientations in a cirrus cloud. Several parametrizations were developed for visible and infrared wavelengths (Magono, 1966; Labonnote et al., 2000; Yang et al., 2001, 2005; Baum et al., 2005b, 2011; Baran and Labonnote, 2007; Baran et al., 2009; Baran, 2012; Baran et al., 2013). For cirrus cases 1 to 8 we employ the aggregate ice crystal model (Yang et al., 2001, 2005) with a monodisperse distribution used in the IIR retrieval algorithm (Garnier et al., 2013) that provides an extinction coefficient, a single-scattering albedo and an asymmetry factor (Yang et al., 2001, 2005). Note that Dubuisson et al. (2008) have shown that the IIR thermal infrared channels are weakly sensitive to the ice crystal shape and almost insensitive to the size distribution. The IIR retrieval algorithm uses three ice crystal shapes (Garnier et al., 2012, 2013), namely a solid column, aggregate and plate. The phase functions of these

Table 1. Mean cloud properties of the cirrus generated by 3DCLOUD. "CTA" corresponds to the cirrus-top altitude; "OP" corresponds to the optical properties parametrization; "Yal" represents the model of ice crystals developed by Yang et al. (2001, 2005) for aggregates ice crystals; and "Bal" represents the parametrization of ice crystals' optical properties developed by Baran et al. (2009), Baran (2012), and Baran et al. (2013); τ_c is the cloud mean optical thickness; σ_τ is the cloud standard deviation of the optical thickness estimated from the optical thickness of the subpixels at the scale of $100\,\mathrm{m} \times 100\,\mathrm{m}$; ρ_τ is the cloud heterogeneity parameter defined as the ratio of σ_τ by τ_c; and D_{eff} is the ice crystal effective diameter.

Cirrus	CTA (km)	τ_c	σ_τ	ρ_τ	D_{eff} (μm)	OP
1	7.97	0.45	0.32	0.7	9.95	Yal
2	7.97	0.90	0.63	0.7	9.95	Yal
3	7.97	1.80	1.26	0.7	9.95	Yal
4	7.97	1.80	1.98	1.1	9.95	Yal
5	7.97	1.80	2.70	1.5	9.95	Yal
6	7.97	1.80	1.26	0.7	20.09	Yal
7	7.97	1.80	1.26	0.7	40.58	Yal
8	11.06	0.90	0.63	0.7	9.95	Yal
CII-1	11.06	0.41	0.32	0.77	heterogeneous	Bal
CII-2	11.06	0.81	0.62	0.77	heterogeneous	Bal
CII-3	11.06	0.90	0.63	0.70	9.95	Yal

particles are relatively smooth in the thermal infrared with a small forward peak (asymmetry factor g usually below 0.9) and can be approximated by the Henyey–Greenstein phase function. While this assumption is certainly problematic for irregular crystal shapes, as shown by Baum et al. (2005a, b), we use the Henyey–Greenstein phase function to remain consistent with the official IIR retrieval algorithm (Garnier et al., 2012, 2013). For these cirrus cases, the optical properties are constant over the entire cloud.

In order to generate 3-D and heterogeneous cloud optical properties fields for the CII-1 and CII-2 cirrus cases, we used the parametrization of Baran et al. (2009, 2013) and Baran (2012). This parametrization, derived from in situ measurements of more than 20 000 particle-size distributions (Field et al., 2005, 2007), gives the optical coefficients as a function of IWC and temperature.

2.3 TOA brightness temperature simulations

TOA brightness temperatures in the three IIR thermal infrared channels (8.65, 10.60 and 12.05 μm) are simulated with the 3DMCPOL code developed in the visible range by Cornet et al. (2010) and extended to the infrared range by Fauchez et al. (2014). 3DMCPOL is a forward Monte Carlo algorithm using the local estimate method (Marshak and Davis, 2005; Mayer, 2009) and is able to simulate radiances and brightness temperatures from the visible to the infrared range, including the polarization. The atmosphere is subdivided in voxels (3-D pixels), with a constant horizontal size ($\mathrm{d}x$, $\mathrm{d}y$) and a variable vertical size ($\mathrm{d}z$). Each voxel is described by the extinction coefficient σ_e, the single-

scattering albedo ϖ_0, the phase function and the cloud temperature T_c.

3-D BT are first simulated at $100\,\mathrm{m} \times 100\,\mathrm{m}$ spatial resolution and are then averaged to the IIR spatial resolution of $1\,\mathrm{km} \times 1\,\mathrm{km}$ ($\mathrm{BT}_{1\,\mathrm{km}}^{3\text{-}D}$). 1-D BT are obtained by averaging the optical property field to $1\,\mathrm{km} \times 1\,\mathrm{km}$ spatial resolution before simulating the BT ($\mathrm{BT}_{1\,\mathrm{km}}^{1\text{-}D}$).

Note that the statistical uncertainty of these simulations is below 0.5 K, which is less than the IIR accuracy of about 1 K. Comparisons between 3DMCPOL statistical uncertainty, IIR accuracy and heterogeneity effects can be found in Fauchez et al. (2014) (Figs. 8 and 10) for the same cloud scenes. This statistical uncertainty is reached by simulating between 5 and 10 billion photons for each case.

2.4 Retrieval algorithms of cloud parameters

Two related algorithms are used to retrieve cloud products: the split-window technique (Inoue, 1985; Parol et al., 1991; Dubuisson et al., 2008) to retrieve the effective diameter and an algorithm similar to the IIR operational algorithm to retrieve the effective emissivity and the effective optical thickness.

In the thermal infrared atmospheric window, the SWT is one of the most used methods to retrieve the effective diameter and the cloud optical thickness using the difference of brightness temperatures between two thermal infrared channels (Parol et al., 1991; Radel et al., 2003; Dubuisson et al., 2008; Garnier et al., 2012, 2013). Figure 2 shows brightness temperature difference (BTD) for varying optical thickness (0–50 at 12.05 μm) and eight effective diameters (D_{eff}) as a function of the 12.05 μm BT (BT_{12}). Each "arch" corresponds to a single effective size, with BTD decreasing with increasing particle size and optical thickness decreasing along each arch from opaque cloud (low BT) to clear sky (high BT). It is evident that the sensitivity of the SWT to large particles ($D_{\mathrm{eff}} > 40\,\mu$m) is weak, one of the main disadvantages of this method that can only accurately determine the effective size of particles smaller than approximately 40 μm for cirrus clouds with an optical thickness approximately between 0.5 and 3 (Dubuisson et al., 2008; Sourdeval et al., 2012). Dubuisson et al. (2008) also show that the SWT retrieval accuracy for ice crystal effective diameter is between 10 and 25 % and for the optical thickness is about 10 %. We note that the amplitude of the BTD_{8-10} arches is significantly smaller than the two others because its sensitivity to D_{eff} is weaker. Consequently, this channel pair will not be used in this study.

Similar to the SWT, the IIR operational algorithm (Garnier et al., 2012) uses radiance differences between channels, though in a different way. Intermediate products (effective emissivity, effective optical thickness and microphysical indices) are computed to retrieve the ice crystal effective diameter and shape. The effective emissivity refers to the contribution of scattering in the retrieved emissivity, especially for small ice crystals in the band at 8.65 μm. One of the major advantages of using the effective emissivity is its independence of cloud-top altitude or geometrical thickness, contrary to the brightness temperature differences used in the SWT. The effective emissivity, $\varepsilon_{\text{eff},k}$, for the channel k is defined as

$$\varepsilon_{\text{eff},k} = [R_k - R_{k,\text{BG}}]/[B_k(T_c, Z_c) - R_{k,\text{BG}}], \tag{2}$$

where R_k is the measured (or simulated) radiance in the channel k, $R_{k,\text{BG}}$ is the measured (or simulated) radiance at TOA for clear sky and $B_k(T_c, Z_c)$ is the radiance of an opaque cloud (black body) located at the centroid altitude Z_c and at the centroid temperature T_c, provided by the GEOS-5 model (Rienecker et al., 2008). The layer centroid altitude is a weighted average altitude based on the attenuated backscattered intensity of the LIDAR signal at 532 nm (Vaughan et al., 2009). Note that, in this study, we set the centroid altitude to the geometrical middle of the cloud.

The effective optical thickness $\tau_{\text{eff},k}$ is then calculated as

$$\tau_{\text{eff},k} = -\ln(1 - \varepsilon_{\text{eff},k}). \tag{3}$$

From $\tau_{\text{eff},k}$, the microphysical indices $\text{MI}^{12/8}$ and $\text{MI}^{12/10}$ are defined as the ratio of $\tau_{\text{eff},k}$ between 12.05 and 8.65 μm channels and 12.05 and 10.60 μm channels, respectively:

$$\text{MI}_{12/8} = \tau_{\text{eff},12}/\tau_{\text{eff},8}, \quad \text{MI}_{12/10} = \tau_{\text{eff},12}/\tau_{\text{eff},10}. \tag{4}$$

These microphysical indices strongly depend on the microphysical and optical properties of the cloud layer, namely the effective diameter and shape of the ice crystals. From a look-up table (LUT) of the microphysical indices as a function of the effective diameter and shape precalculated by the FASDOM code (Dubuisson et al., 2005), two values of effective diameters ($D_{\text{eff},1\,\text{km}}(10, 8)$ and $D_{\text{eff},1\,\text{km}}(12, 8)$) are obtained for each particle shape (aggregates, plates and solid columns) considered in the IIR retrieval algorithm. The shape corresponding to the smallest difference between the two $D_{\text{eff},1\,\text{km}}$ is selected. For the computation of optical properties, the IIR operational algorithm uses the Yang et al. (2001, 2005) model with a monomodal effective diameter distribution.

The uncertainty of the retrieval algorithm was checked by comparing optical properties retrieved from simulated radiances with the optical properties used as input in the radiative transfer. For this, we perform a 1-D retrieval from 1-D-simulated radiances. The algorithm uncertainties are less than 2 % for effective diameters retrieved with the SWT (test not shown here) and 4 % for effective optical thickness retrieved with the algorithm similar to the IIR operational algorithm (test not shown here).

3 Impact of cirrus heterogeneities on the retrieved parameters

In this section, we present the heterogeneity effects on the retrieved products at the 1 km IIR spatial resolution as a

function of different cloud optical properties (i.e., optical thickness, effective diameters, extinction coefficients, single-scattering albedo and asymmetry factor) and microphysical (IWC) properties, cirrus-top altitude and geometrical thickness. The heterogeneity effects on the retrieved parameters are assessed by using the difference between products retrieved from modeled 3-D ($\text{BT}_{1\,\text{km}}^{3\text{-D}}$) and 1-D ($\text{BT}_{1\,\text{km}}^{1\text{-D}}$) 1 km brightness temperatures.

In order to estimate the heterogeneity effects on the retrieved cloud products, we define the following errors due to cloud heterogeneities:

$$\Delta\varepsilon_{\text{eff}} = \varepsilon_{\text{eff}}^{3\text{-D}} - \varepsilon_{\text{eff}}^{1\text{-D}} \tag{5}$$

for effective emissivities calculated by the Eq. (2);

$$\Delta\tau_{\text{eff}} = \tau_{\text{eff}}^{3\text{-D}} - \tau_{\text{eff}}^{1\text{-D}} \tag{6}$$

for effective optical thicknesses calculated by Eq. (3);

$$\Delta\text{MI}^{12/8} = \text{MI}^{3\text{-D},12/8} - \text{MI}^{1\text{-D},12/8} \text{ and}$$
$$\Delta\text{MI}^{12/10} = \text{MI}^{3\text{-D},12/10} - \text{MI}^{1\text{-D},12/10} \tag{7}$$

for microphysical indices calculated from Eq. (4);

$$\Delta D_{\text{eff},1\,\text{km}} = D_{\text{eff},1\,\text{km}}^{3\text{-D}} - D_{\text{eff},1\,\text{km}}^{1\text{-D}} \tag{8}$$

for ice crystal effective diameter retrieved with the SWT.

The "3-D" exponent corresponds to optical properties retrieved from $\text{BT}_{1\,\text{km}}^{3\text{-D}}$ and the "1-D" exponent corresponds to those retrieved from $\text{BT}_{1\,\text{km}}^{1\text{-D}}$. $D_{\text{eff},1\,\text{km}}^{1\text{-D}}$ corresponds either to the effective diameter used in the radiative transfer simulation when it is known (cirrus cases 1 to 8 and CII-3) or to the effective diameter retrieved from $\text{BT}_{1\,\text{km}}^{1\text{-D}}$ when the ice crystal effective diameters used in the radiative transfer simulation are unknown (cirrus cases CII-1 and CII-2).

Heterogeneity impacts due to the optical thickness variability are discussed in Sect. 3.1 and those due to optical and microphysical property variabilities in Sect. 3.2.

3.1 Heterogeneity impacts due to the optical thickness variability

Fauchez et al. (2014) show that $\text{BT}_{1\,\text{km}}^{3\text{-D}}$ are larger than $\text{BT}_{1\,\text{km}}^{1\text{-D}}$ and that their difference is well correlated with the standard deviation of the optical thickness inside the 1 km $\times$ 1 km observation pixel $\sigma_{\tau_{1\,\text{km}}}$. This brightness temperature difference is due to the plane-parallel approximation caused by the non-linearity of the relationship between brightness temperature and optical thickness. The impact of the PPA bias ($|\text{BT}_{1\,\text{km}}^{3\text{-D}} - \text{BT}_{1\,\text{km}}^{1\text{-D}}|$) is greater (in absolute value) for highly absorbing bands because the increase of cloud absorption leads to a larger brightness temperature contrast between the cirrus top and the clear sky atmosphere and, thus, to a stronger averaging effect. Figure 3 illustrates how cirrus heterogeneities affect the retrieval of the effective

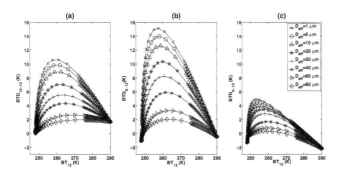

Figure 2. Brightness temperatures differences (BTD) as a function of the $12.05\,\mu m$ brightness temperature (BT_{12}) for eight effective diameters (D_{eff}) and different optical thickness between 0 and 50 at $12.05\,\mu m$: **(a)** BTD_{10-12} between 10.60 and $12.05\,\mu m$ channels, **(b)** BTD_{8-12} between 8.65 and $12.05\,\mu m$ channels and **(c)** BTD_{8-10} between 8.65 and $10.60\,\mu m$ channels.

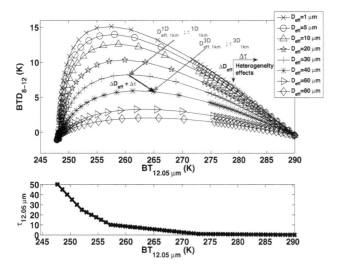

Figure 3. Top panel: brightness temperature differences between 8.65 and $12.05\,\mu m$ (BTD_{12-8}) as a function of the brightness temperature at $12.05\,\mu m$ (BT_{12}). The red arrow shows an example of effective diameter, $D_{eff,\,1\,km}^{1-D}$, and optical thickness, $\tau_{1\,km}^{1-D}$, retrieved in 1-D without heterogeneity effects, and the blue arrow shows the corresponding effective diameter, $D_{eff,\,1\,km}^{3-D}$, and optical thickness, $\tau_{1\,km}^{3-D}$, retrieved with heterogeneity effects. Each point of the arches corresponds to an optical thickness represented in the bottom panel, with $\tau_{12.05\,\mu m}$ as the optical thickness at $12.05\,\mu m$. Using the plane-parallel approximation (PPA) leads to an overestimation of the effective diameter and to underestimation of the optical thickness, respectively, compared to a 3-D retrieval.

diameter and the optical thickness. The tip of the red arrow represents the BTD and BT values obtained with a homogeneous cloud with $D_{eff,\,1\,km}^{1-D}$ and $\tau_{1\,km}^{1-D}$. Using 3-D radiative transfer inside a heterogeneous cloud with the same mean properties, we obtained the BTD and BT values represented by the tip of the blue arrow in Fig. 3. As heterogeneity effects are larger at the $12.05\,\mu m$ channel than at the $8.65\,\mu m$

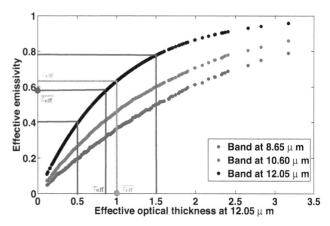

Figure 4. Variation of the effective emissivity as a function of the effective optical thickness at $12.05\,\mu m$, estimated in 1-D at the spatial resolution of 100 m for the three IIR channels and for cloudy pixels belonging to cirrus cases 1 to 5. τ_{eff} represents the effective optical thickness corresponding to the averaged effective emissivity $\overline{\varepsilon_{eff}}$, $\overline{\tau_{eff}}$ represents the averaged effective optical thickness and ε_{eff} is its corresponding effective emissivity. Brown and green lines show the effective emissivity and effective optical thickness values on the x axis and y axis, respectively, corresponding to a particular point on, for instance, the black arch. The mathematical formulation of the PPA is expressed by the Jensen inequality $\overline{\varepsilon_{eff}} < \varepsilon_{eff}(\overline{\tau_{eff}})$.

channel, brightness temperature differences (BTD_{8-12}), first simulated at the 100 m spatial resolution and then averaged to the 1 km IIR spatial resolution, are smaller than those retrieved from radiances directly simulated at 1 km spatial resolution. Consequently, as effective diameters increase with the decrease of BTD, the retrieved $D_{eff,\,1\,km}^{3-D}$ is larger than the mean value $D_{eff,\,1\,km}^{1-D}$ and the retrieved optical thicknesses $\tau_{1\,km}^{3-D}$ is smaller than the mean optical thickness $\tau_{1\,km}^{1-D}$.

In addition, Fig. 4 shows the effective emissivity as a function of the effective optical thickness estimated at 100 m spatial resolution. The relationship between effective emissivities and effective optical thickness is nonlinear, as it is between brightness temperatures and optical thickness. Because of the PPA bias, the average effective emissivity is smaller than the effective emissivity of the average of the effective optical thickness $\overline{\tau_{eff}}$. Similar to brightness temperatures, effective emissivities and effective optical thickness retrieved from radiances, first simulated at 100 m spatial resolution of and then averaged to the IIR spatial resolution of 1 km, are smaller than those retrieved from radiances directly simulated at the spatial resolution of 1 km.

Figure 5 presents $\Delta\varepsilon_{eff}$ (a, b and c) and $\Delta\tau_{eff}$ (d, e and f) as a function of the standard deviation of the optical thickness inside the 1 km × 1 km observation pixel ($\sigma_{\tau_{1\,km}}$) for cirrus cases 1 to 5 and for 8.65, 10.60 and $12.05\,\mu m$ channels, respectively. We notice, first of all, that $\Delta\varepsilon_{eff}$ and $\Delta\tau_{eff}$ are correlated with $\sigma_{\tau_{1\,km}}$ at more than 94 % except for cirrus case 1 at $8.65\,\mu m$, where the horizontal transport smooths the slight

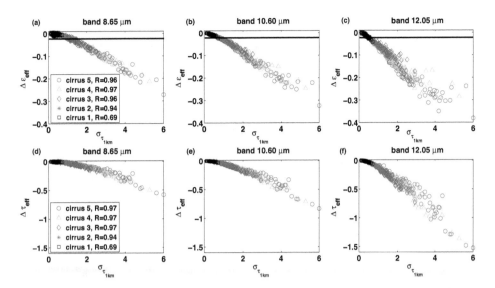

Figure 5. Errors on the effective emissivity $\Delta\varepsilon_{\text{eff}}$ (**a, b** and **c**) and on the effective optical thickness $\Delta\tau_{\text{eff}}$ (**d, e** and **f**) at 8.65, 10.60 and 12.05 μm, respectively, as a function of the optical thickness standard deviation, $\sigma_{\tau_{1\,\text{km}}}$, for cirrus cases 1 ($\tau_c = 0.45$, $\rho_\tau = 0.7$), 2 ($\tau_c = 0.90$, $\rho_\tau = 0.7$), 3 ($\tau_c = 1.80$, $\rho_\tau = 0.7$), 4 ($\tau_c = 1.80$, $\rho_\tau = 1.1$) and 5 ($\tau_c = 1.80$, $\rho_\tau = 1.5$) with $D_{\text{eff}} = 9.95\,\mu$m for the five cirrus. The black lines correspond to the IIR operational algorithm uncertainty on the effective emissivity.

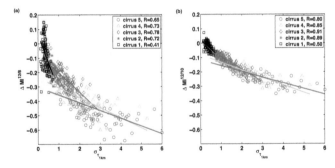

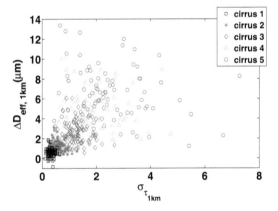

Figure 6. Microphysical index differences $\Delta\text{MI}^{12/8}$ (**a**) and $\Delta\text{MI}^{12/10}$ (**b**) as a function of the standard deviation of the optical thickness, $\sigma_{\tau_{1\,\text{km}}}$, for cirrus cases 1 ($\tau_c = 0.45$, $\rho_\tau = 0.7$), 2 ($\tau_c = 0.90$, $\rho_\tau = 0.7$), 3 ($\tau_c = 1.80$, $\rho_\tau = 0.7$), 4 ($\tau_c = 1.80$, $\rho_\tau = 1.1$) and 5 ($\tau_c = 1.80$, $\rho_\tau = 1.5$) with $D_{\text{eff}} = 9.95\,\mu$m for the five cirrus. R represents the correlation coefficient between ΔMI and $\sigma_{\tau_{1\,\text{km}}}$.

Figure 7. Errors on the retrieved effective diameter $\Delta D_{\text{eff},1\,\text{km}}$ as a function of the standard deviation of the optical thickness, $\sigma_{\tau_{1\,\text{km}}}$, for cirrus cases 1 ($\tau_c = 0.45$, $\rho_\tau = 0.7$), 2 ($\tau_c = 0.90$, $\rho_\tau = 0.7$), 3 ($\tau_c = 1.80$, $\rho_\tau = 0.7$), 4 ($\tau_c = 1.80$, $\rho_\tau = 1.1$) and 5 ($\tau_c = 1.80$, $\rho_\tau = 1.5$) with $D_{\text{eff}} = 9.95\,\mu$m for the five cirrus. Effective diameters are estimated using the split-window technique.

heterogeneity of the radiative field. $\Delta\varepsilon_{\text{eff}}$ and $\Delta\tau_{\text{eff}}$ are negative, meaning that the 3-D effective emissivities and effective optical thickness are smaller than those in 1-D. Indeed, as explained by Fauchez et al. (2014), heterogeneity effects lead to an increase of radiances or brightness temperatures. As radiances decrease with the cloud extinction, larger radiances lead then to smaller cloud effective emissivity and effective optical thickness. In addition, $\Delta\varepsilon_{\text{eff}}$ and $\Delta\tau_{\text{eff}}$ are shown to depend on the wavelength. For example, at $\sigma_{\tau_{1\,\text{km}}} = 1$, $\Delta\varepsilon_{\text{eff}}$ is equal to -0.01 at 8.65 μm, -0.03 at 10.60 μm and -0.05 at 12.05 μm. This is due to the increase of absorption from 8.65 to 12.05 μm that leads to an increase of the contrast between cloud and clear sky pixels, and thus to an increase of the PPA bias. For comparison, Garnier et al. (2012) have shown that

the effective emissivity error due to the retrieval method is about 0.03 for the 12.05 μm band (black lines of the Fig. 5) assuming a 1 K clear sky atmosphere temperature uncertainty for an ocean scene. This uncertainty is smaller than the average error due to cloud heterogeneity $\overline{\Delta\varepsilon_{\text{eff}}}$. We can note that, at $\sigma_{\tau_{1\,\text{km}}} \sim 1$, $\Delta\varepsilon_{\text{eff}}$ is equal to or larger than 0.03 for the 10.60 and 12.05 μm bands. $\sigma_{\tau_{1\,\text{km}}} \sim 1$ corresponds also to the limit where the heterogeneity effects on brightness temperatures become larger than the IIR instrumental accuracy of 1 K (Fauchez et al., 2014).

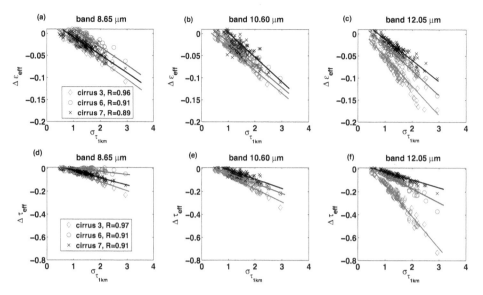

Figure 8. Errors on the effective emissivity $\Delta\varepsilon_{\text{eff}}$ (**a**, **b** and **c**) and on the effective optical thickness $\Delta\tau_{\text{eff}}$ (**d**, **e** and **f**) at 8.65, 10.60 and 12.05 μm, respectively, as a function of the optical thickness standard deviation, $\sigma_{\tau_{1 \text{ km}}}$, for three identical cirrus fields but for different ice crystal effective diameters: cirrus cases 3 ($D_{\text{eff}} = 9.95\,\mu$m), 6 ($D_{\text{eff}} = 20.09\,\mu$m) and 7 ($D_{\text{eff}} = 40.58\,\mu$m), with $\tau_c = 1.80$ and $\rho_\tau = 0.7$ for the three cirrus. R represents the correlation coefficient between $\Delta\varepsilon_{\text{eff}}$ (**a**, **b** and **c**) and $\sigma_{\tau_{1 \text{ km}}}$ and between $\Delta\tau_{\text{eff}}$ (**d**, **e** and **f**) and $\sigma_{\tau_{1 \text{ km}}}$.

Figure 6a and b show the error on the microphysical indices $\Delta MI^{12/8}$ and $\Delta MI^{12/10}$, respectively, as a function of $\sigma_{\tau_{1 \text{ km}}}$ for cirrus cases 1 to 5. First note that the errors on the two microphysical indices are on average negative, except again for cirrus case 1 for $\Delta MI^{12/8}$, and they increase with the cirrus mean optical thickness (from cirrus cases 1 to 3) and heterogeneity parameter (from cirrus cases 3 to 5). The correlation with $\sigma_{\tau_{1 \text{ km}}}$ is better for $\Delta MI^{12/10}$ than for $\Delta MI^{12/8}$. Again, the strongest scattering in the band at 8.65 μm tends to smooth the radiative field heterogeneities and, therefore, to degrade the correlation between $\Delta MI^{12/8}$ and $\sigma_{\tau_{1 \text{ km}}}$. $\Delta MI^{12/8}$ is, on average, larger than $\Delta MI^{12/10}$ because the difference of absorption and effective emissivity is significantly greater for the 12.05 μm / 8.65 μm pair than for 12.05 μm / 10.60 μm. Effective diameters of ice crystals are estimated from the microphysical indices using a LUT and are thus also impacted by heterogeneity effects. As $\Delta MI = MI^{3\text{-D}} - MI^{1\text{-D}}$ is negative, the impact of cloud heterogeneities leads to an underestimation of the microphysical indices. This underestimation leads then to an overestimation of the retrieved effective diameters (smaller microphysical indices correspond to larger effective diameters).

Using the SWT, we are also able to simulate the impact of cirrus heterogeneities on the retrieved effective diameters of ice crystals. In Fig. 7, we plot the error on the effective diameter error $\Delta D_{\text{eff}, 1 \text{ km}}$, due to heterogeneities, as a function of $\sigma_{\tau_{1 \text{ km}}}$ for cirrus cases 1 to 5. We see that $\Delta D_{\text{eff}, 1 \text{ km}}$ is positive and generally increases with the cirrus mean optical thickness (from cirrus cases 1 to 3) and the heterogeneity parameter ρ_τ (from cirrus cases 3 to 5). Indeed, $\sigma_{\tau_{1 \text{ km}}}$ generally increases with τ_c and ρ_τ, as expected.

Figure 8 is the same as Fig. 5, except for different effective diameters: $D_{\text{eff}} = 9.95\,\mu$m, $D_{\text{eff}} = 20.09\,\mu$m and $D_{\text{eff}} = 40.58\,\mu$m (cirrus cases 3, 6 and 7, respectively). Here $\Delta\varepsilon_{\text{eff}}$ and $\Delta\tau_{\text{eff}}$ decrease with increasing D_{eff} (except at the 8.65 μm band where ϖ_0 increases between $D_{\text{eff}} = 9.95$ and 20.09 μm). Indeed, ϖ_0 increases with D_{eff} (except at the 8.65 μm band) and leads to a decrease of the absorption and, thus, of the PPA bias. The impact of the effective diameter on $\Delta\varepsilon_{\text{eff}}$ and $\Delta\tau_{\text{eff}}$ is particularly marked for the 12.05 μm band where the absorption of ice crystals decreases strongly between $D_{\text{eff}} = 9.95\,\mu$m and $D_{\text{eff}} = 40.58\,\mu$m (cirrus cases 3 and 7, respectively).

In addition, we estimated the heterogeneity effects on the retrieved ice crystal effective diameters ($\Delta D_{\text{eff}, 1 \text{ km}}$) for the three D_{eff}. On average, $\Delta D_{\text{eff}, 1 \text{ km}} \sim +3\,\mu$m for cirrus case 3 ($D_{\text{eff}} = 9.95\,\mu$m) and 6 ($D_{\text{eff}} = 20.09\,\mu$m). Thus, there is no a significant increase of heterogeneity effects on retrieved effective diameters between these two effective sizes. For cirrus case 7 ($D_{\text{eff}} = 40.58\,\mu$m), there is no real effect ($\Delta D_{\text{eff}, 1 \text{ km}} \sim \pm 0\,\mu$m) due to the saturation of the SWT. Indeed, as noted above, effective diameters close to 40 μm lead to weak brightness temperature differences. This is illustrated in Fig. 2, where the amplitude of arches, and thus the sensitivity, decreases with the increase of the effective diameter.

3.2 Heterogeneity effects due to optical and microphysical property variabilities

As presented in Sect. 2.2, we use the parametrization developed by Baran et al. (2009, 2013) and Baran (2012) to sim-

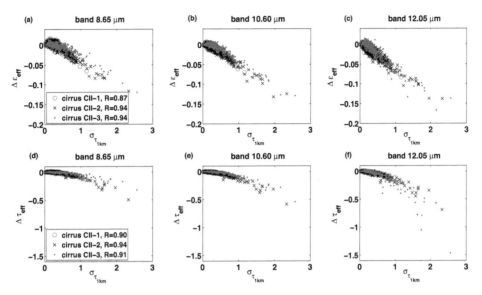

Figure 9. Errors on the effective emissivity $\Delta\varepsilon_{\text{eff}}$ (**a**, **b** and **c**) and on the effective optical thickness $\Delta\tau_{\text{eff}}$ (**d**, **e** and **f**) at 8.65, 10.60 and 12.05 μm, respectively, as a function of the optical thickness standard deviation, $\sigma_{\tau_{1\,\text{km}}}$, for cirrus cases CII-1, CII-2 and CII-3.

ulate a 3-D heterogeneous cloud optical property field from the 3-D distribution of the IWC and temperature. IWC values measured during the CIRCLE-2 campaign are coupled with a MLS temperature profile to generate a realistic 3-D optical property field for simulating of the CII-1 and CII-2 cirrus cases. In addition, to compare with the previous cirrus cases, the cirrus case CII-3 was generated from the CIRCLE-2 cloud field using optical properties identical to cirrus case 8.

Figure 9 shows the impact of cirrus heterogeneities on the retrieved effective emissivity and on the effective optical thickness as a function of the standard deviation of the optical thickness, $\sigma_{\tau_{1\,\text{km}}}$, for cirrus cases CII-1, CII-2 and CII-3. $\Delta\varepsilon_{\text{eff}}$ and $\Delta\tau_{\text{eff}}$ are similar for the three cirrus cases, although some slight differences are evident as a function of the wavelength. Indeed, at 8.65 μm, $\Delta\varepsilon_{\text{eff}}$ and $\Delta\tau_{\text{eff}}$ are smaller for the CII-3 cirrus case than for the two others cirrus cases. At 10.60 μm, this difference is close to 0. At 12.05 μm, $\Delta\varepsilon_{\text{eff}}$ and $\Delta\tau_{\text{eff}}$ are larger for the CII-3 cirrus than for CII-1 and CII-2 cirrus. This effect is due to the variability of the optical properties for the CII-1 and CII-2 cirrus. Indeed, cirrus case CII-3 contains only aggregate crystals of effective diameter $D_{\text{eff}} = 9.95\,\mu$m resulting from the model of Yang et al. (2001, 2005), while cirrus cases CII-1 and CII-2 contain crystal of various sizes. For CII-3 cirrus, small crystals have a single-scattering albedo maximum at 8.65 μm, leading to a lower PPA bias. At 12.05 μm, small particles are more absorbing and the PPA bias is larger. For the CII-1 cirrus, corresponding to the cirrus observed during the CIRCLE-2 campaign, the average effective emissivity error is within the limit of the method sensibility (Garnier et al., 2012) of about 0.03 in absolute value.

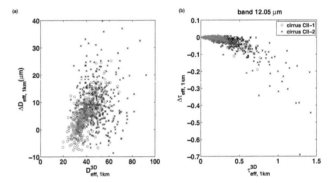

Figure 10. (**a**) Errors on the retrieved effective diameter $\Delta D_{\text{eff},\,1\,\text{km}}$ as a function of the effective diameter, $D_{\text{eff},\,1\,\text{km}}^{3\text{-D}}$, and (**b**) error on the effective optical thickness, $\Delta\tau_{\text{eff},\,1\,\text{km}}$, as a function of the effective optical thickness, $\tau_{\text{eff},\,1\,\text{km}}^{3\text{-D}}$, for cirrus CII-1 and CII-2.

To study heterogeneity effects on the retrieved ice crystals' effective diameters for the CII-1 and CII-2 cirrus, we compare effective diameters $D_{\text{eff},\,1\,\text{km}}^{3\text{-D}}$ and $D_{\text{eff},\,1\,\text{km}}^{1\text{-D}}$ retrieved from $\text{BT}_{1\,\text{km}}^{3\text{-D}}$ and $\text{BT}_{1\,\text{km}}^{1\text{-D}}$, respectively. $D_{\text{eff},\,1\,\text{km}}^{3\text{-D}}$ and $\tau_{\text{eff},\,1\,\text{km}}^{3\text{-D}}$ represent the cloud optical properties resulting from a 3-D radiative transfer simulation through a heterogeneous atmosphere ($\text{BT}_{1\,\text{km}}^{3\text{-D}}$). The differences $\Delta D_{\text{eff},\,1\,\text{km}} = D_{\text{eff},\,1\,\text{km}}^{3\text{-D}} - D_{\text{eff},\,1\,\text{km}}^{1\text{-D}}$ and $\Delta\tau_{\text{eff},\,1\,\text{km}} = \tau_{\text{eff},\,1\,\text{km}}^{3\text{-D}} - \tau_{\text{eff},\,1\,\text{km}}^{1\text{-D}}$ correspond, therefore, to the heterogeneity effects on the retrieval of $D_{\text{eff},\,1\,\text{km}}^{3\text{-D}}$ and $\tau_{\text{eff},\,1\,\text{km}}^{3\text{-D}}$. For these two cirrus, the optical properties are heterogeneous. Therefore, Fig. 10a shows $\Delta D_{\text{eff},\,1\,\text{km}}$ as a function of $D_{\text{eff},\,1\,\text{km}}^{3\text{-D}}$ and Fig. 10b shows $\Delta\tau_{\text{eff},\,1\,\text{km}}$ as a function of $\tau_{\text{eff},\,1\,\text{km}}^{3\text{-D}}$. We see that $\Delta D_{\text{eff},\,1\,\text{km}}$ and $\Delta\tau_{\text{eff},\,1\,\text{km}}$ increase, in absolute value, with

Table 2. $\overline{D_{\mathrm{eff,IIR}}}$ and $\overline{\tau_{\mathrm{IIR}}}$: averaged effective diameter and optical thickness, respectively, retrieved by IIR on 25 May 2007 during the CIRCLE-2 campaign; $\overline{D^{3\text{-D}}_{\mathrm{eff,1\ km}}}$; and $\overline{\tau^{3\text{-D}}_{1\ km}}$: averaged effective diameter and optical thickness, respectively, retrieved for CII-1 and CII-2 cirrus and $\overline{\Delta D^{3\text{-D}}_{\mathrm{eff,1\ km}}}$ and $\overline{\Delta \tau^{3\text{-D}}_{\mathrm{eff,1\ km}}}$: averaged errors on the effective diameter and optical thickness, respectively, due to cloud heterogeneities in absolute value and in percentage.

Cirrus	$\overline{D_{\mathrm{eff,\ IIR}}}$ (μm)	$\overline{D^{3\text{-D}}_{\mathrm{eff,\ 1\ km}}}$ (μm)	$\overline{\Delta D^{3\text{-D}}_{\mathrm{eff,\ 1\ km}}}$ (μm)	$\overline{\tau_{\mathrm{IIR}}}$	$\overline{\tau^{3\text{-D}}_{\mathrm{eff,\ 1\ km}}}$	$\overline{\Delta\tau^{3\text{-D}}_{\mathrm{eff,\ 1\ km}}}$
CII-1	44.2	38.9	5.1 (13 %)	0.41	0.40	-0.02 (-5 %)
CII-2	–	48.7	9.7 (20 %)	–	0.74	-0.05 (-7 %)

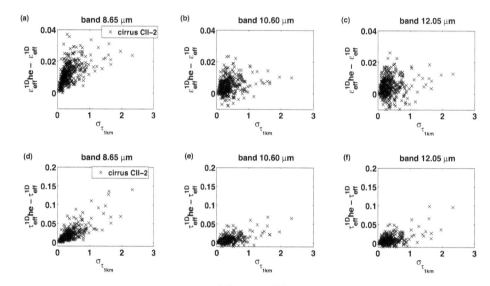

Figure 11. Effective emissivity differences (**a, b, c**) between $\varepsilon^{1\text{-D}}_{\mathrm{eff}}$he and $\varepsilon^{1\text{-D}}_{\mathrm{eff}}$ and effective optical thickness differences (**d, e, f**) between $\tau^{1\text{-D}}_{\mathrm{eff}}$he and $\tau^{1\text{-D}}_{\mathrm{eff}}$ retrieved from radiances calculated in the case of vertically heterogeneous and vertically homogeneous cloudy columns, respectively, as a function of the standard deviation of the optical thickness $\sigma_{\tau_{1\ km}}$ for cirrus CII-2 for bands at 8.65, 10.60 and 12.05 μm, respectively.

$D^{3\text{-D}}_{\mathrm{eff,\ 1\ km}}$ and $\tau^{3\text{-D}}_{\mathrm{eff,\ 1\ km}}$, respectively. Table 2 summarizes the optical properties retrieved by IIR during CIRCLE-2 and those retrieved from our simulations as well as the estimated heterogeneity effects. First of all, for the CII-1 cirrus possessing the characteristics of the cirrus observed during the CIRCLE-2 campaign, the average value of the retrieved effective diameter ($\overline{D^{3\text{-D}}_{\mathrm{eff,\ 1\ km}}} \sim 38.9\,\mu$m) and the mean effective optical thickness ($\overline{\tau^{3\text{-D}}_{\mathrm{eff,\ 1\ km}}} \sim 0.40$ at 12.05 μm) are close to those retrieved from the IIR measurements along the CALIOP/CALIPSO track ($D_{\mathrm{eff,\ IIR}} = 44.2\,\mu$m and $\tau_{\mathrm{IIR}} = 0.41$ without underlying liquid water cloud). Thus, there is a good agreement between optical properties retrieved by the IIR operational algorithm during the CIRCLE-2 campaign and those retrieved with our simulations. The mean error due to heterogeneity effects is approximately 5.1 μm (13 %) for retrieved effective diameter and approximately -0.02 (5 %) for effective optical thickness. On average, these relative errors due to heterogeneity effects are, thus, weak compared to the uncertainty estimate of Dubuisson et al. (2008) for the IIR retrieval (10 to 25 % for $D_{\mathrm{eff,\ 1\ km}}$ and 10 % for $\tau_{\mathrm{eff,\ 1\ km}}$).

However, at the observation pixel scale, some values can reach more than 40 % for effective diameter and 15 % for effective optical thickness, which is quite significant. Furthermore, errors due to cloud heterogeneities increase with the IWC or the cirrus mean optical thickness; the cirrus CII-2 case, for instance, with IWC twice as large as cirrus CII-1, has $\overline{\Delta D_{\mathrm{eff,\ 1\ km}}} \sim 9.7\,\mu$m (20 %) and $\overline{\Delta \tau_{\mathrm{eff,\ 1\ km}}} \sim -0.05$ (7 %).

3.3 Influence of the vertical variability of optical properties

To find the influence of the vertical variability of cirrus optical properties (σ_{e}, ϖ_0 and g) on the retrieval errors, we compare cloud products retrieved from $\mathrm{BT}^{1\text{-D}}_{1\ km}$ with vertically heterogeneous columns with those retrieved for vertically homogeneous columns obtained after a vertical averaging of the IWC, for the CII-2 cirrus case.

Figures 11 shows the effects of the vertical heterogeneity of the optical properties on the effective emissivity (a, b and c) and on the effective optical thickness (d, e and f).

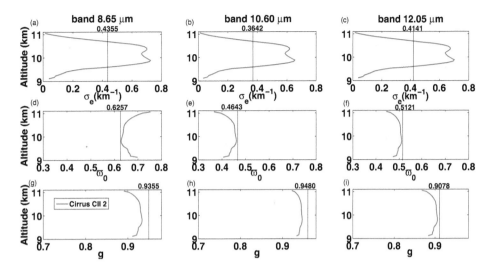

Figure 12. (a), **(b)** and **(c)**: vertical variation of the mean extinction coefficient σ_e; **(d)**, **(e)** and **(f)**: vertical variation of the mean single-scattering albedo ϖ_0; **(g)**, **(h)** and **(i)**: vertical variation of the asymmetry factor g for the three IIR channels at 8.65, 10.60 μm and 12.05 for cirrus CII-2. Vertical black lines correspond to the mean value of the optical coefficient obtained after vertical averaging of the IWC.

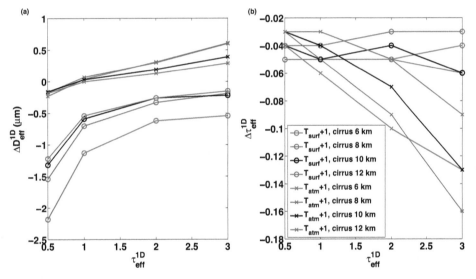

Figure 13. Errors on the **(a)** retrieved effective diameter, $\Delta D_{\mathrm{eff}}^{1\text{-}D}$, and **(b)** on the effective optical thickness, $\Delta \tau_{\mathrm{eff}}^{1\text{-}D}$, as a function of $\tau_{\mathrm{eff}}^{1\text{-}D}$ due to an error of $+1$ K on the surface temperature ($T_{\mathrm{surf}}+1$) and $+1$ K on the atmospheric temperature profile ($T_{\mathrm{atm}}+1$) for cirrus with top altitudes of 6, 8, 10 and 12 km.

Here $\varepsilon_{\mathrm{eff}}^{1\text{-}D}$he and $\tau_{\mathrm{eff}}^{1\text{-}D}$he are estimated from vertically heterogeneous cloudy columns and $\varepsilon_{\mathrm{eff}}^{1\text{-}D}$ and $\tau_{\mathrm{eff}}^{1\text{-}D}$ from vertically homogeneous cloudy columns, as a function of the standard deviation of the optical thickness $\sigma_{\tau_{1\,\mathrm{km}}}$ for the three IIR channels. Differences between retrieved products estimated from vertically heterogeneous and homogeneous cloudy columns are significantly weaker than those due to 3-D heterogeneities (horizontal and vertical heterogeneities). Furthermore, contrary to the 3-D heterogeneity effects $\Delta\varepsilon_{\mathrm{eff}}$ and $\Delta\tau_{\mathrm{eff}}$, the differences ($\varepsilon_{\mathrm{eff}}^{1\text{-}D}he-\varepsilon_{\mathrm{eff}}^{1\text{-}D}$) and ($\tau_{\mathrm{eff}}^{1\text{-}D}he-\tau_{\mathrm{eff}}^{1\text{-}D}$) are positive. These effects are particular to our simulations, where vertical heterogeneities tend thus to smooth the hor-

izontal heterogeneity effects. These observations can be explained with Fig. 12, which shows the vertical profiles of the optical properties of cirrus CII-2 in the vertically heterogeneous case (red curves) and the vertically homogeneous case (black lines) after vertical averaging of the IWC and temperature using the parametrization of Baran et al. (2009, 2013) and Baran (2012). In this way, values of the vertically homogeneous case are different from the average of the optical coefficients of the vertically heterogeneous case: ϖ_0 of the vertically homogeneous case is larger than the vertical averaging of the heterogeneous case for the 10.60 and 12.05 μm bands. In addition, the asymmetry parameter g of the verti-

Table 3. Averaged errors on the retrieved cirrus optical properties due to the 3-D cloud heterogeneity for three ice crystal effective diameters (columns 3, 4 and 5); the vertical heterogeneity of optical properties (column 6) with 1Dvhe and 1Dvho representing the -1-D radiative transfer with vertically heterogeneous and homogeneous columns, respectively; an incertitude of 1 K of the surface temperature (column 7) and the temperature atmospheric profile (column 8); and the IIR retrieval uncertainty (Dubuisson et al., 2008, column 9). $\overline{\Delta D_{\mathrm{eff}_{1\,\mathrm{km}}}}$ and $\overline{|\Delta D_{\mathrm{eff}_{1\,\mathrm{km}}}|}$ correspond to the absolute error in micrometers and to the relative error in percent, respectively, on the retrieval of the effective diameter; $\overline{\Delta \tau_{1\,\mathrm{km}}}$ and $\overline{|\Delta \tau_{1\,\mathrm{km}}|}$ correspond to the absolute and relative error in percent, respectively, on the retrieval of the optical thickness.

		Horizontal heterogeneity effects as a function of D_{eff}			Vertical heterogeneity (1Dvhe – 1Dvho)	surface 1 K	atmosphere 1 K	IIR uncertainty
$\sigma_{\tau_{1\,\mathrm{km}}}$	$D_{\mathrm{eff}}\ (\mu\mathrm{m})$	40.58	20.09	9.95	48.7	9.95	9.95	–
1	$\overline{\Delta D_{\mathrm{eff, 1\,km}}}\ (\mu\mathrm{m})$	−0.5	2.0	2.5	2	1	0.2	–
	$\overline{|\Delta D_{\mathrm{eff, 1\,km}}|}\ (\%)$	~ 1	~ 10	~ 25	~ 4	~ 10	~ 2	~ 10 to ~ 25
	$\overline{\Delta \tau_{\mathrm{eff, 1\,km}}}$	−0.02	−0.10	−0.20	0.03	0.04	0.08	–
	$\overline{|\Delta \tau_{\mathrm{eff, 1\,km}}|}\ (\%)$	~ 1	~ 6	~ 12	~ 4	~ 2	~ 4	~ 10
2	$\overline{\Delta D_{\mathrm{eff, 1\,km}}}\ (\mu\mathrm{m})$	1	3	3	2	–	–	–
	$\overline{|\Delta D_{\mathrm{eff, 1\,km}}|}\ (\%)$	~ 3	~ 15	~ 40	~ 4	–	–	~ 10 to ~ 25
	$\overline{\Delta \tau_{\mathrm{eff, 1\,km}}}$	−0.10	−0.20	−0.50	0.10	–	–	–
	$\overline{|\Delta \tau_{\mathrm{eff, 1\,km}}|}\ (\%)$	~ 6	~ 12	~ 28	~ 12	–	–	~ 10

cally homogeneous case is larger than the average of the vertically heterogeneous case in the three bands. Consequently, the cirrus is less absorbent in the vertically homogeneous case and thus the effective emissivities and effective optical thicknesses are weaker. This vertical variability of optical properties, for the cirrus CII-2 case, impact the retrieval of the effective diameter of, on average, $4\,\mu\mathrm{m}$ (figure not presented here).

Note that effects of the vertical variability are discussed here for the structure of the cirrus observed during the CIRCLE-2 campaign. Effects could be different for other cirrus structure but they are not discussed here. For example, for old cirrus, sedimentation processes could be much larger, increasing differences between the cloud top and base. The impact of the vertical variability on cloud properties retrieved from satellite observations could thus be larger.

4 Other sources of uncertainty

We show in the previous sections that heterogeneity effects can be an important source of errors in the retrieved optical properties. To compare its importance on the retrieved cloud parameters with regard to other possible error sources for IIR measurements, we test the impact of a 1 K uncertainty in the surface temperature and in the atmospheric temperature profile measurements, an error that corresponds to that estimated by Garnier et al. (2012). Figure 13a and b show the error in the retrieved effective diameter ($\Delta D_{\mathrm{eff}}^{1\text{-}\mathrm{D}}$) and in the retrieved effective optical thickness ($\Delta \tau_{\mathrm{eff}}^{1\text{-}\mathrm{D}}$), respectively, as a function of $\tau_{\mathrm{eff}}^{1\text{-}\mathrm{D}}$ for cirrus with a top altitude of 6, 8, 10 and 12 km. The retrieval of the effective diameter and optical thickness is performed using the SWT on 1-D radiative trans-

fer simulations. We can see that $\Delta D_{\mathrm{eff}}^{1\text{-}\mathrm{D}}$ are less than $2.5\,\mu\mathrm{m}$ (25 %) and $\Delta \tau_{\mathrm{eff}}^{1\text{-}\mathrm{D}}$ less than 0.16 (5 %). By comparison, these errors are in the IIR retrieval uncertainty of 10–25 % for D_{eff} and about 10 % for τ_{eff} (Dubuisson et al., 2008). In addition, they are significantly smaller than those due to cloud heterogeneity effects (more than 50 % for D_{eff} and 10 to 15 % for τ_{eff}). In Fig. 13a, it is evident that increasing cirrus optical thickness or cloud-top altitude decreases the effective diameter retrieval error due to a 1 K surface-temperature uncertainty. In Fig. 13b, the effective diameter retrieval error due to atmospheric temperature profile uncertainty increases with increasing optical thickness because cloud emissivity also increases.

Fauchez et al. (2014) show that cloud-top altitude and geometrical thickness significantly influence the heterogeneity effects because the brightness temperature contrast between the surface and the cloud top increases with increasing cloud-top altitude and decreases with increasing vertical extension for a constant cloud top (as the cloud base is closer to the surface). For retrieved cloud products estimated with an algorithm similar to the IIR operational algorithm, the effective emissivity is independent of the cloud altitude and geometrical thickness; thus the impacts of altitude and geometrical thickness on the retrieval are weak.

5 Summary and conclusions

In this paper, we discussed the impact of cirrus heterogeneity effects in the retrieval of cloud parameters from thermal infrared radiometric measurements from space. We have focused on the IIR radiometer for which the operational algorithm estimates the cirrus effective emissivity, the effective optical thickness and the ice crystal effective diameter of the observation pixel. We show that errors due to the cirrus heterogeneity effects on the effective emissivity and the effective optical thickness are well correlated to the subpixel optical thickness standard deviation $\sigma_{\tau_{1 \, \mathrm{km}}}$ and generally increase with increasing optical thickness $\tau_{1 \, \mathrm{km}}$. These errors are greater than the precision of the retrieval method ($\Delta\varepsilon_{\mathrm{eff}} \sim 0.03$) for $\sigma_{\tau_{1 \, \mathrm{km}}} \sim 1$, corresponding also to the value from which heterogeneity effects on brightness temperatures become larger than the IIR instrumental accuracy of 1 K (Fauchez et al., 2014).

Our results are summarized in Table 3. Heterogeneity effects for three effective diameters are compared with the retrieval errors caused by the vertical inhomogeneity of optical properties and with the impact of an error of 1 K, corresponding to the IIR accuracy, on the atmospheric temperature profile and on the surface temperature. Results are shown for pixels with $\sigma_{\tau_{1 \, \mathrm{km}}} = 1$ (medium heterogeneity) and 2 (large heterogeneity). The most important errors in the cloud optical property retrieval concern those due to the subpixel heterogeneity of the optical thickness, in particular for the smallest crystals ($\overline{\Delta D_{\mathrm{eff}, 1 \, \mathrm{km}}} = 2.5 \, \mu m$ ($\sim 25 \, \%$) and $\Delta\tau_{\mathrm{eff} \, 1 \, \mathrm{km}} = -0.20$ ($\sim 12 \, \%$) for $\sigma_{\tau_{1 \, \mathrm{km}}} = 1$). Indeed, the absorption is larger for small crystals and, thus, the PPA bias is greater. For $D_{\mathrm{eff}} = 40.58 \, \mu m$, the ice crystal optical properties in the three IIR channels converge to similar values leading to smaller brightness temperature differences between channels and, thus, to a decrease of the retrieval accuracy. Errors due to the vertical inhomogeneity of the optical properties, an error of 1 K on the surface temperature or atmospheric temperature profile are smaller than the IIR retrieval errors (Dubuisson et al., 2008). Thus, the influence of these parameters appears negligible compared to optical thickness heterogeneity and IIR retrieval uncertainty.

The impacts of cirrus heterogeneities on the retrieved cloud parameters studied in this paper are for a 1 km spatial resolution. These biases could decrease with an increase of the spatial resolution although photon transport effects would increase. Fauchez et al. (2014) estimate that a 250 m spatial resolution could significantly reduce the PPA bias while photon transport effects remain weak. However, heterogeneity effects on the retrieved cloud products at this resolution require further investigation. This study also provides ways to potentially correct the heterogeneity errors using the subpixel measurements to estimate $\sigma_{\tau_{1 \, \mathrm{km}}}$. Furthermore, differences between heterogeneity effects in the visible/near-infrared and thermal infrared ranges for different spatial resolutions also require further investigation to estimate their impact on cloud products retrieved using a combination of the visible and near-infrared/short-wave infrared and infrared retrieval methods, as proposed by Cooper et al. (2007).

Acknowledgements. The authors acknowledge the Centre National de la Recherche Scientifique, the Programme National de Télédétection Spatiale and the Direction Générale de l'Armement for their financial support. We also thank the use of resources provided by the European Grid Infrastructure. For more information, please refer to the EGI-InSPIRE paper (http://go.egi.eu/pdnon).

We thank Anthony Baran for the numerous fruitful discussions on the cirrus optical properties.

We also thank the two reviewers whose remarks largely improved the quality of this article.

Edited by: A. Macke

References

Baran, A. J.: From the single-scattering properties of ice crystals to climate prediction: A way forward, Atmos. Res., 112, 45–69, 2012.

Baran, A. J. and Labonnote, L.-C.: A self-consistent scattering model for cirrus. I: The solar region, Q. J. Roy. Meteorol. Soc., 133, 1899–1912, 2007.

Baran, A. J., Connolly, P. J., and Lee, C.: Testing an ensemble model of cirrus ice crystals using midlatitude in situ estimates of ice water content, volume extinction coefficient and the total solar optical depth., J. Quant Spectrosc. Ra., 110, 1579–1598, 2009.

Baran, A. J., Cotton, R., Furtado, K., Havemann, S., Labonnote, L.-C., Marenco, F., Smith, A., and Thelen, J.-C.: A self-consistent scattering model for cirrus. II: The high and low frequencies, Q. J. Roy. Meteorol. Soc., 140, 1039–1057, 2013.

Baum, B., Heymsfield, A., Yang, P., Platnick, S., King, M., Hu, Y.-X., and Bedka, S.: Bulk scattering models for the remote sensing of ice clouds. Part 1: Microphysical data and models, J. Appl. Meteor., 44, 1885–1895, 2005a.

Baum, B., Yang, P., Heymsfield, A., Platnick, S., King, M., Hu, Y.-X., and Bedka, S.: Bulk scattering properties for the remote sensing of ice clouds. Part II. Narrowband models, J. Appl. Meteor., 44, 1896–1911, 2005b.

Baum, B., Yang, P., Heymsfield, A., Schmitt, C., Xie, Y., Bansemer, A., Hu, Y.-X., and Zhang, Z.: Improvements in shortwave bulk scattering and absorption models for the remote sensing of ice clouds, J. Appl. Meteorol. Clim., 50, 1037–1056, 2011.

Cahalan, R. F., Ridgway, W., Wiscombe, W. J., Bell, T. L., and Snider, J. B.: The Albedo of Fractal Stratocumulus Clouds, J. Atmos. Sci., 51, 2434–2455, 1994.

Carlin, B., Fu, Q., Lohmann, U., Mace, J., Sassen, K., and Comstock, J. M.: High cloud horizontal inhomogeneity and solar albedo bias, J. Climate, 15, 2321–2339, 2002.

Cooper, S. J., L'Ecuyer, T. S., Gabriel, P., Baran, A. J., and Stephens, G. L.: Performance assessment of a five-channel estimation-based ice cloud retrieval scheme for use over the global oceans, J. Geophys. Res.-Atmos., 112, D04207, doi:10.1029/2006JD007122, 2007.

Cornet, C., C-Labonnote, L., and Szczap, F.: Three-dimensional polarized Monte Carlo atmospheric radiative transfer model (3DM-CPOL): 3D effects on polarized visible reflectances of a cirrus cloud, J. Quant. Spectr. Radiat. Trans., 111, 174–186, 2010.

Dubuisson, P., Giraud, V., Chomette, O., Chepfer, H., and Pelon, J.: Fast radiative transfer modeling for infrared imaging radiometry, J. Quant. Spectrosc. Ra., 95, 201–220, 2005.

Dubuisson, P., Giraud, V., Pelon, J., Cadet, B., and Yang, P.: Sensitivity of thermal infrared radiation at the top of the atmosphere and the surface to ice cloud microphysics, J. Appl. Meteorol. Clim., 47, 2545–2560, 2008.

Fauchez, T., Cornet, C., Szczap, F, Dubuisson, P., and Rosambert, T.: Impact of cirrus clouds heterogeneities on top-of-atmosphere thermal infrared radiation, Atmos. Chem. Phys., 14, 5599–5615, doi:10.5194/acp-14-5599-2014, 2014.

Field, P. R., Hogan, R. J., Brown, P. R. A., Illingworth, A. J., Choularton, T. W., and Cotton, R. J.: Parametrization of ice-particle size distributions for mid-latitude stratiform cloud, Q. J. Roy. Meteorol. Soc., 131, 1997–2017, 2005.

Field, P. R., Heymsfield, A., and Bansemer, A.: Snow size distribution parameterization for midlatitude and tropical ice clouds, J. Atmos. Sci., 64, 4346–4365, 2007.

Garnier, A., Pelon, J., Dubuisson, P., Faivre, M., Chomette, O., Pascal, N., and Kratz, D. P.: Retrieval of Cloud Properties Using CALIPSO Imaging Infrared Radiometer. Part I: Effective Emissivity and Optical Depth, J. Appl. Meteorol. Clim., 51, 1407–1425, 2012.

Garnier, A., Pelon, J., Dubuisson, P., Yang, P., Faivre, M., Chomette, O., Pascal, N., Lucker, P., and Tim, M.: Retrieval of Cloud Properties Using CALIPSO Imaging Infrared Radiometer. Part II: effective diameter and ice water path, J. Appl. Meteorol. Clim., 52, 2582–2599, 2013.

Hogan, R. J. and Kew, S. F.: A 3D stochastic cloud model for investigating the radiative properties of inhomogeneous cirrus clouds, Q. J. Roy. Meteorol. Soc., 131, 2585–2608, 2005.

Inoue, T.: On the temperature and effective emissivity determination of semi-transparent cirrus clouds by bi-spectral measurements in the 10-μ m window region, J. Meteorol. Soc. Jpn., 63, 88–99, 1985.

Katagiri, S., Sekiguchi, M., Hayasaka, T., and Nakajima, T.: Cirrus cloud radiative forcing at the top of atmosphere using the nighttime global distribution with the microphysical parameters derived from AVHRR, AIP Conf., Proc. 1531, 704–707, 2013.

Kato, S. and Marshak, A.: Solar zenith and viewing geometry-dependent errors in satellite retrieved cloud optical thickness: Marine stratocumulus case, J. Geophys. Res.-Atmos., 114, D01202, doi:10.1029/2008JD010579, 2009.

Labonnote, C. L., Brogniez, G., Doutriaux-Boucher, M., Buriez, J.-C., Gayet, J.-F., and Chepfer, H.: Modeling of light scattering in cirrus clouds with inhomogeneous hexagonal monocrystals. Comparison with in-situ and ADEOS-POLDER measurements, Geophys. Res. Lett., 27, 113–116, 2000.

Liou, K. N.: Influence of Cirrus Clouds on Weather and Climate Processes: A Global Perspective, Mon. Weather Rev., 114, 1167–1199, 1986.

Lynch, D. K., Sassen, K., Starr, D. O. C., and Stephens, G.: Cirrus, Oxford University Press, USA, 2002.

Magono, C.: Meteorological Classification of Natural Snow Crystals, Hokkaido University, 1966.

Marshak, A. and Davis, A. A. B.: 3D radiative transfer in cloudy atmospheres, Physics of Earth and Space Environments Series, Springer-Verlag Berlin Heidelberg, 2005.

Mayer, B.: Radiative transfer in the cloudy atmosphere, Eur. Phys. J. Conferences, 1, 75–99, 2009.

Mioche, G., Josset, D., Gaye, J. F., Pelon, J., Garnier, A., Minikin, A., and Schwarzenboeck, A.: Validation of the CALIPSO-CALIOP extinction coefficients from in situ observations in midlatitude cirrus clouds during the CIRCLE-2 experiment, J. Geophys. Res.-Atmos., 115, D00H25, doi:10.1029/2009JD012376, 2010.

Nakajima, T. and King, M. D.: Determination of the optical thickness and effective particle radius of clouds from reflected solar radiation measurements. Part I: Theory, J. Atmos. Sci., 47, 1878–1893, 1990.

Parol, F., Buriez, J. C., Brogniez, G., and Fouquart, Y.: Information Content of AVHRR Channels 4 and 5 with Respect to the Effective Radius of Cirrus Cloud Particles, J. Appl. Meteorol., 30, 973–984, 1991.

Radel, G., Stubenrauch, C. J., Holz, R., and Mitchell, D. L.: Retrieval of effective ice crystal size in the infrared: Sensitivity study and global measurements from TIROS-N Operational Vertical Sounder, J. Geophys. Res.-Atmos., 108, 4281, doi:10.1029/2002JD002801, 2003.

Rienecker, M. M., Suarez, M. J., Todling, R., Bacmeister, J., Takacs, L., Liu, H.-C., Gu, W., Sienkiewicz, M., Koster, R. D., Gelaro, R., Stajner, I., and Nielsen, J. E.: Technical Report Series on Global Modeling and Data Assimilation, Technical Report, 27, 2008.

Sassen, K. and Cho, B. S.: Subvisual-Thin Cirrus Lidar Dataset for Satellite Verification and Climatological Research, J. Appl. Meteorol., 31, 1275–1285, 1992.

Sourdeval, O., Brogniez, G., Pelon, J., Labonnote, C. L., Dubuisson, P., Parol, F., Josset, D., Garnier, A., Faivre, M., and Minikin, A.: Validation of IIR/CALIPSO Level 1 Measurements by Comparison with Collocated Airborne Observations during CIRCLE-2 and Biscay '08 Campaigns, J. Atmos. Oceanic Technol., 29, 653–667, 2012.

Starr, D. O. and Cox, S. K.: Cirrus Clouds. Part II: Numerical Experiments on the Formation and Maintenance of Cirrus, J. Atmos. Sci., 42, 2682–2694, 1985.

Stephens, G. L.: Radiative Properties of Cirrus Clouds in the Infrared Region, J. Atmos. Sci., 37, 435–446, 1980.

Stephens, G. L., Gabriel, P. M., and Tsay, S.-C.: Statistical radiative transport in one-dimensional media and its application to the terrestrial atmosphere, Transport Theory Stat. Phys., 20, 139–175, 1991.

Szczap, F., Isaka, H., Saute, M., Guillemet, B., and Gour, Y.: Inhomogeneity effects of 1D and 2D bounded cascade model clouds on their effective radiative properties, Phys. Chem. Earth Pt. B, 25, 83–89, 2000.

Szczap, F., Gour, Y., Fauchez, T., Cornet, C., Faure, T., Jourdan, O., Penide, G., and Dubuisson, P.: A flexible three-dimensional stratocumulus, cumulus and cirrus cloud generator (3DCLOUD) based on drastically simplified atmospheric equations and the Fourier transform framework, Geosci. Model Dev., 7, 1779–1801, doi:10.5194/gmd-7-1779-2014, 2014.

Varnai, T. and Marshak, A.: Statistical analysis of the uncertainties in cloud optical depth retrievals caused by three-dimensional radiative effects, J. Atmos. Sci., 58, 1540–1548, 2001.

Vaughan, M. A., Powell, K. A., Kuehn, R. E., Young, S. A., Winker, D. M., Hostetler, C. A., Hunt, W. H., Liu, Z., McGill, M. J., and Getzewich, B. J.: Fully Automated Detection of Cloud and

Aerosol Layers in the CALIPSO Lidar Measurements, J. Atmos. Oceanic Technol., 26, 2034–2050, 2009.

Yang, P., Liou, K. N., Wyser, K., and Mitchell, D.: Parameterization of the scattering and absorption properties of individual ice crystals, J. Geophys. Res.-Atmos., 105, 4699–4718, 2000.

Yang, P., Gao, B.-C., Baum, B. A., Hu, Y. X., Wiscombe, W. J., Tsay, S.-C., Winker, D. M., and Nasiri, S. L.: Radiative properties of cirrus clouds in the infrared (8–13 um) spectral region, J. Quant. Spectrosc. Ra., 70, 473–504, 2001.

Yang, P., Wei, H., Huang, H.-L., Baum, B. A., Hu, Y. X., Kattawar, G. W., Mishchenko, M. I., and Fu, Q.: Scattering and absorption property database for nonspherical ice particles in the near-through far-infrared spectral region, Appl. Opt., 44, 5512–5523, 2005.

Zinner, T. and Mayer, B.: Remote sensing of stratocumulus clouds: Uncertainties and biases due to inhomogeneity, J. Geophys. Res.-Atmos., 111, D14209, doi:10.1029/2005JD006955, 2006.

6

An overview of the lightning and atmospheric electricity observations collected in southern France during the HYdrological cycle in Mediterranean EXperiment (HyMeX), Special Observation Period 1

E. Defer[1], J.-P. Pinty[2], S. Coquillat[2], J.-M. Martin[2], S. Prieur[2], S. Soula[2], E. Richard[2], W. Rison[3], P. Krehbiel[3], R. Thomas[3], D. Rodeheffer[3], C. Vergeiner[4], F. Malaterre[5], S. Pedeboy[5], W. Schulz[6], T. Farges[7], L.-J. Gallin[7], P. Ortéga[8], J.-F. Ribaud[9], G. Anderson[10], H.-D. Betz[11], B. Meneux[11], V. Kotroni[12], K. Lagouvardos[12], S. Roos[13], V. Ducrocq[9], O. Roussot[9], L. Labatut[9], and G. Molinié[14]

[1]LERMA, UMR8112, Observatoire de Paris & CNRS, Paris, France
[2]LA, UMR5560, Université de Toulouse & CNRS, Toulouse, France
[3]NMT, Socorro, New Mexico, USA
[4]Institute of High Voltage Engineering and System Performance, TU Graz, Graz, Austria
[5]Météorage, Pau, France
[6]OVE-ALDIS, Vienna, Austria
[7]CEA, DAM, DIF, Arpajon, France
[8]GePaSUD, UPF, Faa'a, Tahiti, French Polynesia
[9]CNRM-GAME, UMR3589, Météo-France & CNRS, Toulouse, France
[10]UK Met Office, Exeter, UK
[11]nowcast, Garching, Germany
[12]NOA, Athens, Greece
[13]Météo France, Nîmes, France
[14]LTHE, Grenoble, France

Correspondence to: E. Defer (eric.defer@obspm.fr)

Abstract. The PEACH project (Projet en Electricité Atmosphérique pour la Campagne HyMeX – the Atmospheric Electricity Project of the HyMeX Program) is the atmospheric electricity component of the Hydrology cycle in the Mediterranean Experiment (HyMeX) experiment and is dedicated to the observation of both lightning activity and electrical state of continental and maritime thunderstorms in the area of the Mediterranean Sea. During the HyMeX SOP1 (Special Observation Period) from 5 September to 6 November 2012, four European operational lightning locating systems (ATDnet, EUCLID, LINET, ZEUS) and the HyMeX lightning mapping array network (HyLMA) were used to locate and characterize the lightning activity over the northwestern Mediterranean at flash, storm and regional scales. Additional research instruments like slow antennas, video cameras, microbarometer and microphone arrays were also operated. All these observations in conjunction with operational/research ground-based and airborne radars, rain gauges and in situ microphysical records are aimed at characterizing and understanding electrically active and highly precipitating events over southeastern France that often lead to severe flash floods. Simulations performed with cloud resolving models like Meso-NH and Weather Research and Forecasting are used to interpret the results and to investigate further the links between dynamics, microphysics, electrification and lightning occurrence. Herein we present an overview of the PEACH project and its different instruments. Examples are discussed to illustrate the comprehensive and unique lightning data set, from radio frequency to acoustics, collected during the SOP1 for lightning phenomenology understanding, instrumentation validation, storm characterization and modeling.

1 Introduction

A lightning flash is the result of an electrical breakdown occurring in an electrically charged cloud. Charged regions inside the cloud are created through electrification processes dominated by ice–ice interactions. Electrical charges are exchanged during rebounding collisions between ice particles of different nature in the presence of supercooled water. This corresponds to the most efficient non-inductive charging process investigated by Takahashi (1978) and Saunders et al. (1991). Laboratory studies have shown that the transfer of electrical charges between ice particles in terms of amount and sign is very complex and depends on the difference of velocity between the two ice particles, temperature and liquid water content. The lighter hydrometeors are transported upward, the heaviest being sustained at lower altitude in the cloud. Combined with cloud dynamics and cloud microphysics, electrification processes lead to dipoles, tripoles and even stacks of charged zones vertically distributed in the thundercloud (Stolzenburg et al., 1998; Rust et al., 2005). Between the charged regions, the ambient electric field can reach very high values, i.e., more than $100\,\mathrm{kV\,m^{-1}}$ (Marshall et al., 2005). However, such an electric field intensity is of 1 order of magnitude lower than the electric field threshold required to breakdown cloud air. Therefore, additional ignition mechanisms have been considered, such as runaway electrons (Gurevich et al., 1992) or hydrometeor interactions present in high electric fields (Crabb and Latham, 1974; Coquillat and Chauzy, 1994; Schroeder et al., 1999; Coquillat et al., 2003). Natural lightning flashes then occur when the ambient electric field exceeds a threshold of a few $\mathrm{kV\,m^{-1}}$. Hence, it is clear that the lightning activity of a thundercloud results from intricate and complex interactions between microphysical, dynamical and electrical processes.

Lightning flashes are usually classified into two groups: intra-cloud (IC) flashes only occur in a cloud, while cloud-to-ground (CG) flashes connect to the ground. Negative (positive) CG flashes lower negative (positive) charge to the ground and exhibit significant electromagnetic radiation when connecting the ground. Negative CG flashes are more frequent than +CG flashes and generally occur with multiple connections to the ground (e.g., Mäkelä et al., 2010; Orville et al., 2011). Positive CG flashes are relatively rare and often composed of a single or very few connections to the ground with higher current than −CG flashes. A natural lightning flash is not a continuous phenomenon but is in fact composed of successive events, also called flash components, with different physical properties in terms of discharge propagation, radio frequency radiation type, current

properties, space and time scales. A lightning flash then constitutes a series of multi-scale physical processes spanning from the electron avalanche to the propagation of discharges over large distances of a few kilometers or more. Each of these sub-processes radiates electromagnetic waves in a wide wavelength spectrum.

Different detection techniques have been developed to detect and locate these processes. They usually operate at specific wavelength ranges and are sensitive to some components of the lightning discharges. For instance, some ground-based or space-borne sensors detect electromagnetic radiation emitted in the very high frequency (VHF) domain (e.g., Proctor, 1981; Shao and Krehbiel, 1996; Jacobson et al., 1999; Krehbiel et al., 2000; Defer et al., 2001; Defer and Laroche, 2009). Other instruments detect the radiation emitted by lightning flashes in the optical wavelength (e.g., Light et al., 2001; Christian et al., 2003) or in the very low frequency / low frequency (VLF / LF) range (e.g., Cummins et al., 1998; Smith et al., 2002; Betz et al., 2008, 2009). But because no technique covers all the physical aspects of a lightning flash, multi-instrumental observations are required to provide the most comprehensive description in order to analyze in great detail the lightning flashes and consequently the whole lightning activity of a thunderstorm.

Lightning flashes can be investigated flash by flash to derive their properties. With appropriate lightning sensors such as VHF lightning mappers, the temporal and spatial evolution of the lightning activity can be related to the characteristics of the parent clouds. The total (IC + CG) flash rate is usually a good indicator of the severity of convective systems (Williams et al., 1999). A sudden increase (decrease) of the flash rate is often associated with a more vigorous convection (storm decay). Flash rates usually increase while the storm is developing because conditions for a significant non-inductive charging process are favorable. Flash rates reach a peak value when the cloud top reaches its maximum altitude and then decrease at the onset of the decaying stage of the parent thundercloud. Links between severe weather phenomena including lightning flashes, tornadoes, hail storms, wind gusts and flash floods have been studied for many years. As IC observations were not widely recorded and disseminated, numerous investigations used CG reports to predict severe weather (e.g., Price et al., 2011; Kohn et al., 2011). However, in the past decade it has been shown that the total lightning activity is a more reliable indicator of severe weather (e.g., MacGorman et al., 1989; Goodman et al., 1988; Williams et al., 1999; Montanyà et al., 2007). Schultz et al. (2011) report that the use of total lightning trends is indeed more effective than CG trends to identify the onset of severe weather, with an average lead time prior to severe weather occurrence higher when total lightning detection is used as compared to CG detection only. Because detection of the electromagnetic lightning signal can be instantaneously recorded, located and analyzed, flash rate, IC / CG ratio, vertical distribution of the lightning activity, flash duration and flash density can be used to identify severe weather in real time but deeper investigations are required.

Having illustrated the potential advantages and the difficulties arising from lightning-storm severity relationships, it is useful to review some available modeling tools to investigate this issue. Among them, 3-D cloud resolving models (CRM) including parameterizations of both electrification mechanisms and lightning discharges are of highest interest. For instance, Mansell et al. (2002) included a very sophisticated lightning flash parameterization in the electrification model of Ziegler et al. (1991). Poeppel (2005) also improved a lightning parameterization in the pioneering model of Helsdon et al. (1987, 2002). Altaratz et al. (2005) concentrated their efforts to test a storm electrification scheme in a regional model (RAMS) but without simulating the lightning flashes, which constitutes by far the most difficult part. More recently, Yair et al. (2010) have developed a method for predicting the potential for lightning activity based on the dynamical and the microphysical fields of the Weather Research and Forecasting (WRF) model. Cloud electrification and discharge processes have also been included recently in the French community model Meso-NH (Molinié et al., 2002; Barthe et al., 2005, 2007; Barthe and Pinty, 2007a, b).

CRMs are the preferred modeling tools to study the sensitivity of the electrical charge structure to the electrification mechanisms (see Barthe et al., 2007a, b). A key challenge in simulating cloud electrification mechanisms is the lack of agreement in the community about the relevance of each of the non-inductive charging diagrams published by Takahashi (1978) and by Saunders et al. (1991). Those diagrams disagree in some way because the protocol of the laboratory experiments was different. As a consequence, changing the non-inductive parameterization rates according to these diagrams deeply modifies the simulated cloud charge structure where regular dipole, inverse dipole or tripole of charge layers can be obtained while keeping the same microphysics and dynamics in the CRMs.

Lightning detection is definitively useful to monitor thunderstorms and to help improve severe weather simulations. Among the open scientific questions related to the electrical activity are the links between microphysics, kinematics and lightning activity, the use of the lightning information in multi-sensor rainfall estimation, and the lightning-flash phenomenology. In the following we describe the rationale for lightning detection to characterize the electrical properties of northwestern Mediterranean storms during a dedicated campaign of the Hydrological cycle in the Mediterranean Experiment (HyMeX) program (Ducrocq et al., 2014). First, the HyMeX project is briefly described in Sect. 2. The scientific questions and the observational strategy of the HyMeX lightning task team, including instruments and models, are described in Sect. 3. Section 4 presents an overview of the observations collected at flash, storm and regional scales. Section 5 then discusses the perspectives by listing out the next steps of the data analysis as well as the data and products made available to the HyMeX community.

2 The HyMeX program

The Mediterranean region is regularly affected by heavy precipitation often causing devastating flash floods. Floods and landslides in the Mediterranean basin cost lives and lead to expensive property damage. Improving the knowledge and forecast of these high-impact weather events is a major objective of the HyMeX program (Ducrocq et al., 2014). As part of this 10-year program, the first Special Observation Period (SOP1) HyMeX field campaign was conducted during 2 months from 5 September 2012 to 6 November 2012 over the northwestern Mediterranean Sea and its coastal regions in France, Italy and Spain. The instrumental and observational strategy of the SOP1 campaign was set up to document and improve the knowledge of atmospheric processes leading to heavy precipitation and flash flooding in that specific Mediterranean region. A large battery of atmospheric research instruments were operated during the SOP1 including, among others, mobile weather Doppler and polarimetric radar, airborne radar, in situ microphysics probes, lidar and rain gauges (Ducrocq et al., 2014; Bousquet et al., 2014). These measurement platforms were deployed at or near super sites where dedicated research instruments are gathered to document specific atmospheric processes (Ducrocq et al., 2014). The research lightning sensors operated during the HyMeX SOP1 were located in the Cévennes–Vivarais (CV) area in southeastern France. Additionally, various operational weather forecasting models were used as detailed in Ducrocq et al. (2014).

The HyMeX program (Ducrocq et al., 2014) and its intensive observation period of autumn 2012 was an interesting opportunity to implement multi-instrumental observations for documenting the various processes related to electrification of thunderstorms in a region prone to thunderstorms and high-precipitation events. This was performed during the PEACH (Projet en Electricité Atmosphérique pour la Campagne HyMeX – the Atmospheric Electricity Project of the HyMeX Program) experiment, the HyMeX atmospheric electricity component, as detailed in the following.

3 The PEACH experiment

Summer electrical activity is predominately located over continental Europe while during the winter the electrical convective clouds are mainly observed over the Mediterranean Sea, as established by climatology based on lightning records (e.g., Holt et al., 2001; Christian et al., 2003; Defer et al., 2005) or on space-based microwave measurements (e.g., Funatsu et al., 2009). Holt et al. (2001) discussed that the largest number of days with thunderstorms over the Mediterranean basin is located near the coasts of Italy and Greece. Based on

3 years of Tropical Rainfall Measurement Mission (TRMM) lightning imaging sensor (LIS) observations, Adamo (2004) reported that the flash rates over the Mediterranean Sea are significantly smaller than those recorded at similar latitudes in the United States. This finding is consistent with the fact that convection and consequently lightning activity are significantly stronger over land than over sea (Christian et al., 2003).

Current geostationary satellites can offer a relatively satisfying revisiting time (15 min) to track the storms but cannot provide sounding information below the cloud top. Space-based passive and active microwave sensors on low-orbit satellite missions such as TRMM (Kummerow et al., 1998) or A-Train (Stephens et al., 2002) only provide a scientifically relevant snapshot of the sampled clouds, but the ability of low-orbit instruments to monitor and track weather systems is very limited. Lightning detection data from ground-based detection networks are available continuously and instantaneously over the continental and maritime Mediterranean area as detailed in the following. Lightning information can monitor severe weather events over continental and maritime Mediterranean region but can also improve weather forecasts with lightning data assimilation (Lagouvardos et al., 2013). However, further scientific investigations are required to document the links between the lightning activity and the dynamical and microphysical properties of the parent clouds in continental and maritime Mediterranean storms. In addition, it is necessary to identify the key parameters derived from operational lightning locating systems (OLLS) records alone or in combination with other meteorological observations to provide suitable proxies for better storm tracking and monitoring over the entire Mediterranean basin.

3.1 Scientific objectives and observational/ modeling strategy

In the frame of the HyMeX program, several international institutes joined their effort to investigate the lightning activity and the electrical state of thunderstorms. This topic is part of the HyMeX working group WG3 dedicated to the study of heavy precipitation events (HPEs), flash floods and floods. The PEACH team, composed of the authors of the present article, identified five observational- and modeling-based scientific objectives in relation to HyMeX goals:

1. Study the relationships between kinematics, microphysics, electrification, aerosols, and lightning occurrence and characteristics;

2. Document the electrification processes and charge structures inside clouds over sea and land, and during sea-to-land and land-to-sea transitions;

3. Promote the use of lightning records for data assimilation, nowcasting and very short-range forecasting applications;

4. Cross-evaluate lightning observations from different OLLSs;

5. Establish climatology of lightning activity over the Mediterranean basin.

The first three scientific objectives exhibit obvious connections to WG3 objectives to document and understand thunderstorms leading to HPEs and flash floods and to explore the pertinence of lightning detection in conjunction (or not) with operational weather observations to improve monitoring and forecasting of the storm activity. The fourth objective focuses on the intercomparison of OLLS records to objectively evaluate what each OLLS technology reports, as lightning detection (with a quasi-instantaneous data delivery to the users) and geostationary imagery are the only two weather-observing techniques readily available over the full Mediterranean basin. The fifth objective aims to document long-term series of lightning-based proxies of thunderstorms during the 10-year duration of the HyMeX program but also from more than 2 decades of past lightning data available from some European OLLSs.

The PEACH observational strategy followed the HyMeX observational strategy with SOP (Special Observation Period), EOP (Enhanced Observation Period) and LOP (Long Observation Period) activities. SOP1 activities are mainly described here while EOP and LOP are briefly discussed as they are still underway at the time of this writing. The SOP1 PEACH strategy consisted of deploying relevant instrumentation from September to November 2012 in key locations together with instruments operated by other HyMeX teams with common temporal and spatial coverage over the CV domain. First, OLLSs with continuous, good-quality coverage of the Mediterranean were identified. Then a total-lightning detection system was considered and a portable lightning mapping array (HyLMA) was selected. Electric field mills (EFMs), slow antennas (SLAs) and induction rings (INRs) were also listed as key instruments for characterizing the ambient electric field, the change of the electric field induced by the lightning occurrence and the electrical charges carried by raindrops at ground level, respectively. Finally, in order to increase the scientific returns, additional research field instruments were operated, including a mobile optical camera combined with electric field measurement (VFRS, video and field record system), microbarometer and microphone arrays (MBA and MPA, respectively) and transient luminous event cameras (Fullekrug et al., 2013). The PEACH project also includes two cloud resolving models, Meso-NH (with its electrification and lightning scheme) and WRF.

As discussed in Duffourg and Ducrocq (2011) and Ducrocq et al. (2014), the southeastern part of France has previously experienced heavy precipitation with devastating flash floods, floods and landslides. The PEACH observational setup in conjunction with the other HyMeX research and operational instrumentation aims at documenting the lightning activity existing, or not, in those heavy-precipitation systems. The HyLMA observations combined with the OLLS records provide the required accurate description of the lightning activity (e.g., flash rate, flash density, IC / CG ratio, vertical and horizontal flash development) to investigate its relationships with the dynamical and microphysical cloud properties in combination with ground-based and airborne radars and in situ measurements. Such an investigation is the basis for developing new lightning-based tools for nowcasting and very short-range forecasting applications. In addition, the HyLMA observations, whether in conjunction or not with ground-based electric field measurements, help to investigate the temporal and spatial evolution of the charge structures inside the clouds, over sea and land, as deduced from the properties of the VHF signal radiated by the different flash components. The capability to map with HyLMA the 3-D structure of the lightning flashes, as well as the regions of electrical charges in the thunderclouds, allows the validation of lightning/electrification schemes implemented in numerical cloud resolving models and the investigation of new lightning data assimilation schemes. Finally, to establish a solid climatology of lightning activity over the Mediterranean basin from more than 2 decades of OLLS records, the study of concurrent HyLMA, OLLSs and VFRS records is required not only to access the actual performances of the OLLSs but also to determine precisely the flash components that OLLSs record with the perspective of a better operational use of OLLS observations.

As a result, the HyMeX SOP1 experiment is probably the first ambitious field experiment in Europe to offer such comprehensive descriptions of lightning activity and of its parent clouds over a mountainous area from the early stage to the decaying phase of the sampled electrical storms. Note that a battery of ground-based and airborne research radars in conjunction with the operational network of Météo-France provided a detailed description of the thunderclouds as detailed in Bousquet et al. (2014). Other instruments were deployed as listed in Ducrocq et al. (2014). In this article we give some examples only of atmospheric electricity observations. Several studies are underway on the electrical properties of thunderstorms relative to cloud properties like cloud structure, microphysics and rain patterns, as derived from radar and satellite observations and in situ measurements.

3.2   Research instruments deployed during the SOP1

3.2.1   The HyMeX lightning mapping array

A 12-station lightning mapping array (Rison et al., 1999; Thomas et al., 2004) was deployed in the HyMeX SOP1 area from spring to autumn 2012 (Fig. 1). The HyLMA stations, located in radio-frequency-quiet (RF-quiet), mainly rural areas, were solar powered and used broadband cell phone modems for communications. Each HyLMA station recorded the arrival times and amplitudes of the peaks of impulsive VHF sources, recording at most one peak in every 80 μs in-

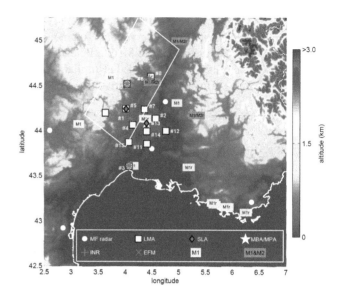

Figure 1. Locations of PEACH instrumental sites (see Table 1 for details on site locations). M1 markers indicate VFRS locations while M2 markers indicate the few locations where a second video camera was operated at the same site; sites where VFRS recorded actual lightning flashes are labeled with an extra letter "r". The Cévennes–Vivarais domain is also delimited by the white polygon.

terval. Locations of impulsive VHF sources were determined by correlating the arrival times for the same event at multiple stations (Thomas et al., 2004). Every minute, a subset of the raw data (the peak in every 400 μs interval) was transferred to a central computer for real-time processing and display. The full data were retrieved at the end of the project for detailed post-processing.

An LMA locates the strongest VHF source in every 80 μs interval. Because negative leaders radiate much more strongly than positive leaders and negative and positive leaders typically propagate at the same time, an LMA primarily locates lightning channels from negative leaders. In particular, an LMA rarely detects the positive leaders from positive cloud-to-ground strokes.

The HyLMA detected all lightning over the array with a location accuracy of about 10 m horizontally and 30 m vertically (Thomas et al., 2004). The HyLMA located much of the lightning outside of the array, with increasingly large location errors (< 1 km at 200 km range) out to a distance of about 300 km from the array center. In order to locate a source, at least six stations must have line of sight to that source. The lines of sight of most of the stations to low-altitude lightning channels outside of the array were blocked by the mountainous terrain in southeastern France, so the LMA typically detected only the higher altitude lightning channels outside the array.

3.2.2 Slow antennas

Two solar-powered slow antennas were deployed to measure the electrostatic field changes from lightning in the SOP1 area. One SLA was deployed a few tens of meters from the MBA/MPA (see Sect. 3.2.3) near the Uzès airfield, and the second was deployed near the HyLMA station at La Grande-Combe airfield. Each SLA consisted of an inverted flat-plate antenna connected to a charge amplifier with a 10 s decay constant. The output of the charge amplifier was digitized at a rate of 50 000 samples per second with a 24 bit A/D converter synchronized to a local GPS receiver, and the data were recorded continuously on SD cards.

3.2.3 The microbarometer and microphone arrays

The CEA (Commissariat à l'Energie Atomique et aux Energies Alternatives) team installed two arrays that overlapped each other: an MBA and MPA. The MBA was composed of four MB2005 microbarometers arranged in an equilateral triangle with sides of about 500 m long with one at the barycenter of the triangle, while the MPA was composed of four microphones arranged in an equilateral triangle with sides of about 52 m long with one at the barycenter of the triangle. The MBA and MPA barycenters were localized at the same place.

Each sensor measures the pressure fluctuation relative to the absolute pressure. The MB2005 microbarometer has a sensitivity of a few millipascals through a band pass of 0.01–27 Hz. This sensor is used in most of the infrasound stations of the international monitoring system of the Comprehensive Nuclear-Test-Ban Treaty Organization (www.ctbto.org). The microphone is an encapsulated BK4196 microphone. Its sensitivity is about 10 mPa through a band pass of 0.1–70 Hz. In order to minimize the noise due to surface wind effects, each sensor is connected to a noise-reducing system equipped with multi-inlet ports (eight for the microbarometers and four for the microphones) that significantly improve the detection capability above 1 Hz. To further reduce the wind noise, microbarometers were installed under vegetative cover (i.e., pine forest).

The signal from these sensors was digitalized at 50 Hz for the MBA and 500 Hz for the MPA. The dating was GPS tagged. Data were stored on a hard disk. No remote access was possible during the SOP1. To avoid power blackouts, each measurement point was supplied with seven batteries. Those batteries needed to be recharged in the middle of the campaign, meaning that the MBA and MPA were unavailable from 9 to 12 October.

The data from each sensor of the arrays were compared using cross-correlation analysis of the waves recorded. The azimuth and the trace velocity were calculated for each detected event when a signal was coherent over the array. Using the time of the lightning discharge and these parameters, a 3-D location of acoustic sources generated by the thunder was

possible (e.g., MacGorman et al., 1981; Farges and Blanc, 2010; Arechiga et al., 2011; Gallin, 2014). Gravity waves generated by thunderstorms (Blanc et al., 2014) could also be monitored by MBA. When a convective system goes over an array, a large pressure variation was measured.

3.2.4 Electric field mills

The surface electrostatic field can be used to detect the presence of charge overhead within a cloud. This parameter is generally measured with a field mill and the value obtained can be very variable according to the sensor shape and location, the relief of the measurement site, the nature of the environment, etc. The field value and its evolution must be interpreted very carefully due to the variety of sources of charge: the cloud charge, the space charge layer that can develop above ground from corona effect on the ground irregularities and the charge carried by the rainfall (Standler and Winn, 1979; Chauzy and Soula, 1987; Soula et al., 2003). However, the electric field evolution can be used to identify discontinuities due to the lightning flashes, which can be related to the flashes detected by location systems (Soula and Georgis, 2013).

The field mills used at three of the stations were Previstorm models from Ingesco Company and were initially used by Montanya et al. (2009). The measurement head is oriented downward to avoid rain disturbances and is fixed at the top of a 1 m mast that reinforces the electrostatic field on the measuring electrode. The measuring head of the fourth field mill was orientated upward and flush to the ground thanks to a hole dug in the ground. The field mills were calibrated to zero by using a shielding and by considering the fair weather conditions that correspond to the theoretical value of 130 V m^{-1}. The data from each sensor were recorded with a time resolution of 1 s. This time resolution readily revealed the major discontinuities in the electrostatic field caused by the lightning flashes without the distracting effects of much faster individual processes within a flash. The polarity of the field was positive when the field points upward and the electric field was created by negative charge overhead.

3.2.5 Induction rings

The electric charge carried by raindrops can easily be detected and measured by an INR, a simple apparatus. This sensor consists of a cylindrical electrode (the ring) on the inner surface, where induced electric charges appear by electrostatic influence when a charge raindrop enters the sensor. When the drop leaves the sensor, the induced charges disappear. The cylindrical electrode is connected to an electrometer and the current signal induced by the passing of a charged drop (a bipolar current impulse) is sampled at a rate of 2000 Hz. It is amplified and integrated by an electronic circuitry that directly provides the charge signal. This one appears as a single pulse with amplitude and length propor-

Table 1. Site ID numbers and locations of the PEACH SOP1 instruments. Sites of VFRS records are not indicated here. MF stands for Météo-France; EMA for Ecole des Mines d'Alès.

ID #	Location	Type	Owner	LMA	SLA	MBA/MPA	INR	EFM
1	Alès	Building roof	EMA school				X	X
2	Cadignac	Land	Private	X				
3	Candillargues	Airfield	Local administration	X			X	X
4	Deaux	Airfield	Local administration	X				
5	La Grande-Combe	Airfield	MF/local administration	X	X			
6	Lavilledieu	Building roof	Elementary school				X	X
7	Méjannes-le-Clap	Land	Local administration	X				
8	Mirabel	Land	Private	X				
9	Mont Aigoual	Land	Private	X				
10	Mont Perier	Land	Private	X			X	X
11	Nîmes	Land	MF	X				
12	Pujaut	Airfield	MF	X				
13	Uzès – north	Airfield	Private		X	X		
14	Uzès – south	Land	MF	X				
15	Vic-le-Fesq	Land	Private	X				

tional to the charge and to the velocity of the drop, respectively. The actual charge is deduced from the calibration of the sensor. If the drop collides with the induction cylinder, the pulse signal exhibits a slow exponential decay (MacGorman and Rust, 1998) that is easily recognizable in the data post-processing. In this case, the raindrop charge that is fully transferred to the induction cylinder is determined by a specific calibration. The charge measurement sensitivity ranges from about ± 2 to ± 400 pC. Furthermore, the charge signal duration at mid-height can be used to determine the size of the charged raindrops, providing the relationship between size and fall velocity is a function of the actual temperature and pressure (Beard, 1976).

Such a measurement provides key information on the electric charge carried by the rain at the ground to validate numerical modeling. It documents the spectrum of charged drops and helps deduce the proportion of charged drops within the whole drop population by comparing its spectrum with the one measured by a disdrometer. Four INRs were built and operated during the SOP1, mainly along the south–north axis at the foothills of the Massif Central where most high-precipitation events occur. Unfortunately, only a few events passed above the sensors and, in these rare cases, the main electronic component of the INRs suffered a malfunction that was not detected during the laboratory tests, so no valuable INR data are available for the SOP1.

3.2.6 Video and field recording system

The VFRS instrument is a transportable system used to measure electric fields and to record high-speed videos at various locations. The calibrated E field measurement consists of a flat-plate antenna, an integrator-amplifier, a fiber optic link

and a digitizer. The bandwidth of the E field measurement ranged from about 350 Hz to about 1 MHz. A 12 bit digitizer with a sampling rate of 5 MS s^{-1} was used for data acquisition. The high-speed camera was operated at 200 fps (equivalent to an exposure of 5 ms frame^{-1}), 640×480 pixel and 8 bit grayscale resolution. The GPS clock provided an accurate time stamp for the E field and the video data. The range of the VFRS was mainly dependent on the visibility conditions. At adequate visibility, combined video and electric field data could record flashes with sufficient quality up to 50 km range. The VFRS was transportable with a car and independent of any external power supply. A detailed description of the used VFRS can be found in Schulz et al. (2005) and in Schulz and Saba (2009). For the typical observations during SOP1, the VFRS was operated in the manual trigger mode using an adjustable pre- and post-trigger. To ensure capturing the entire lightning discharge we typically recorded 6 s of data with 2 s of pre-trigger data per observed flash. During some storms (e.g., low-visibility conditions) the VFRS was operated in the continuous recording mode. Due to memory limitations we only recorded the electric fields in continuous recording mode.

All observation days during SOP1 were chosen based on weather forecasts with sufficient thunderstorm risk over the region of interest. As the real situation could be different to the forecast scenario – e.g., location, motion and stage of the storms – the VFRS sometimes had to be moved from the initial site to another one. For each field operation, the lightning activity of the targeted thunderstorm was monitored in real time using EUCLID and HyLMA observations. The VFRS was often deployed at several sites during a typical observation day. An observation day was finished when no more thunderstorms were expected to occur.

3.2.7 Locations and status of the research instruments

Figure 1 presents the locations of the different PEACH instruments operated during SOP1. The HyLMA network consisted of a dense eight-station network more or less centered on Uzès (Gard) with four additional remote stations located on the western side of the CV domain. SLA antennas were deployed in two different locations: one at the center of the HyLMA network, a few tens of meters away from MPA and MBA, and a second one in the hills, a few hundred meters away from La Grande-Combe HyLMA station (Table 1). INR and EFM were installed on the same sites with other HyMeX SOP1 instruments like the rain gauge, videodistrometers and micro rain radar (MRRs; Bousquet et al., 2014). VFRS observations were performed at different locations during the SOP1 according to the forecast and the evolution of the storm activity, with guidance from HyMeX operation center and members of the lightning team. Finally, the four OLLSs continuously covered the entire SOP1 domain.

Table 2 shows the status of the instruments during the SOP1 and after its completion. HyLMA was initially operational with six stations starting on 1 June 2012 and expanded to 11 stations starting early August 2012. The 12th HyLMA station was online early September 2012. Low time-resolution ($400\,\mu s$ time window) HyLMA lightning observations were delivered in real time during the SOP1 through wireless communication and displayed on the HyMeX operation center website as well as on a dedicated server at NMT. The full HyLMA data were reprocessed after the completion of the SOP1 campaign and only high-temporal resolution HyLMA data are used in the analysis and distributed to the HyMeX community. Additionally, ATDnet, EUCLID and ZEUS observations were also delivered in real time to the HyMeX operation center.

3.3 Operational lightning locating systems

3.3.1 ATDnet

The UK Met Office VLF ATDnet (Arrival Time Differencing NETwork) lightning location network takes advantage of the long propagation paths of VLF sferics (12.5–14.9 kHz) emitted by lightning discharges that propagate over the horizon via interactions with the ionosphere (Gaffard et al., 2008). The ATDnet network consists of 11 that regularly contribute to the "operational network", plus sensors distributed further afield. The waveforms of VLF sferics received at the ATDnet sensors are transmitted to a central processor in Exeter, where the waveforms are compared in order to estimate arrival time differences. These arrival time differences are compared with theoretical arrival time differences for different locations in order to estimate the most likely source location. Current ATDnet processing requires four ATDnet sensors to detect a lightning stroke in order to be able to calculate a single, unambiguous source location. ATDnet predominantly detects sferics created by CG strokes, as the energy and polarization of Sferics created by CG return strokes can travel more efficiently in the Earth–ionosphere waveguide and so are more likely to be detected at longer ranges than typical IC discharges. ATDnet location uncertainties within the region enclosed by the network of sensors are on the order of a few kilometers, i.e., suitable for identifying electrically active cells.

3.3.2 EUCLID

The EUCLID network (EUropean Cooperation for LIghtning Detection) is a cooperation of several European lightning detection networks (Austria, Finland, France, Germany, Italy, Norway, Portugal, Slovenia, Spain and Sweden) that operate state-of-the-art lightning sensors. As of August 2009 the EUCLID network employs 137 sensors, 5 LPATS III, 18 LPATS IV, 15 IMPACT, 54 IMPACT ES/ESP, 3 SAFIR and 42 LS7000 sensors (oldest to newest), all operating over the same frequency range (1–350 kHz) with individually calibrated gains and sensitivities. Data from all of these sensors are processed in real time using a single common central processor that also produces daily performance analyses for each of the sensors. This assures that the resulting data are as consistent as possible throughout Europe. In fact, the Europe-wide data produced by EUCLID are frequently of higher quality than the data produced by individual country networks due to the implicit redundancy produced by shared sensor information. Since the beginning of the cooperation, the performance of the EUCLID network has steadily improved, e.g., with improved location algorithms, with newer sensor technology and by adapting sensor positions because of bad sites. The flash/stroke detection efficiency (DE) of the EUCLID network in the south of France was determined to be 90/87 % for negative and 87/84 % for positive discharges but for a time period where a close sensor was out of order (Schulz et al., 2014). Therefore the values should be rated as lower limits of EUCLID DE in this region. The location accuracy was determined to be 256 m but based on 14 strokes only.

3.3.3 LINET

The LINET system is a modern lightning detection network in the VLF/LF domain (5–100 kHz) developed by nowcast GmbH (Betz et al., 2008, 2009). LINET Europe consists of more than 120 sensors placed in 25 countries. Each sensor includes a field antenna, a GPS antenna and a field processor. The field antenna measures the magnetic flux produced by a lightning discharge. The processor evaluates this signal and combines it with the accurate time provided by the GPS antenna. Compact data files are then sent to a central processing unit where the final stroke solutions are generated. Accurate location of strokes requires that the emitted sig-

Table 2. Status of the instruments during HyMeX SOP1.

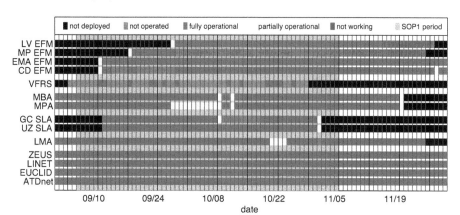

nal is detected by many sensors. Reported strokes are based on reports from at least five sensors. Strokes are located using the time-of-arrival (TOA) method. LINET also detects cloud strokes and can distinguish between CG strokes and IC strokes. Typical baseline of LINET systems are 200 km between adjacent sensors, allowing very good detection efficiency even for very weak strokes (< 10 kA), whereby an average statistical location accuracy of ~ 200 m is achieved. However, in the HyMeX area in southern France the baselines are longer and thus the efficiency is somewhat lower than in most other LINET network areas.

3.3.4 ZEUS

The ZEUS network is a long-range lightning detection system operated by the National Observatory of Athens. The ZEUS system comprises six receivers deployed in Birmingham (UK), Roskilde (Denmark), Iasi (Romania), Larnaca (Cyprus), Athens (Greece) and Lisbon (Portugal), the last being relocated to Mazagón (Spain). ZEUS detects the impulsive radio noise emitted by a lightning strike in the very low frequency (VLF) spectrum between 7 and 15 kHz. At each receiver site an identification algorithm is executed that detects a probable sferics candidate, excludes weak signal and noise and is capable of capturing up to 70 sferics per second. Then the lightning location is retrieved (at the central station) using the arrival time difference technique. Further details on the ZEUS network are given in Kotroni and Lagouvardos (2008). Lagouvardos et al. (2009) have compared the ZEUS system with the LINET system over a major area of central–western Europe, where the latter system presents its major efficiency and accuracy, and found that the location error of ZEUS was 6.8 km and the detection efficiency 25 %. These numbers are also applicable for the SOP1 domain. The authors also found that while ZEUS detects cloud-to-ground lightning it is also capable of detecting strong IC lightning. At this point it should be stated that the statistical analysis showed that ZEUS is able to detect, with high accuracy, the

occurrence of lightning activity although it underdetects the actual number of strokes.

3.4 Instrumentation during EOP and LOP

The only instruments operated so far during EOP and LOP are the OLLSs due to their operational design. For instance, ZEUS observations are continuously delivered in real time to the HyMeX LOP website, while EUCLID and ATDnet produce daily maps of the lightning activity over the Mediterranean basin that are delivered to the HyMeX database. During spring 2014, a network of 12 LMA stations was deployed permanently in Corsica to contribute to the HyMeX LOP efforts in that specific region of the Mediterranean Sea.

3.5 Modeling

3.5.1 The Meso-NH model

The 3-D cloud-resolving mesoscale model Meso-NH (see http://mesonh.aero.obs-mip.fr) contains CELLS, an explicit scheme to simulate the cloud electrification processes (Barthe et al., 2012). This electrical scheme was developed from a one-moment microphysical scheme of Meso-NH to compute the non-inductive charge separation rates (for which several parameterizations are available), the gravitational sedimentation of the charges and the transfer rates as the electrical charges evolve locally according to the microphysical mass transfer rates. The charges are transported by the resolved and turbulent flows. They are carried by cloud droplets, raindrops, pristine ice crystals, snow-aggregates, graupel and two types of positive/negative free ions to close the charge budget. The electric field is computed by inverting the Gauss equation on the model grid (vertical terrain-following coordinate). It is updated at each model time step and also after each flash when several of them are triggered in a single time step. The lightning flashes are treated in a rather coarse way. They are triggered when the electric field reaches the break-even field. A vertically propagating leader is then first initiated to connect the triggering point to the adjacent main layers of charges upwards and downwards. Then the flash propagates horizontally along the layers of charges using a fractal scheme to estimate the number of model grid points reached by the flash path. The flash extension is limited by the geometry of the charged areas and the cloud boundaries. Finally an equal amount of positive and negative charges are partially neutralized at model grid points where an IC flash goes through. In contrast, the CG flashes, detected when the height of the downward tip of the first leader goes 1500 m a.g.l., are polarized since they are not constrained by a neutralization requirement.

3.5.2 The WRF model

The PEACH team has already explored the use of available observational and modeling tools to improve the monitoring, understanding and forecasting of a SOP-like heavy precipitation event over southern France (Lagouvardos et al., 2013). More specifically, the authors applied an assimilation technique that controls the activation of the convective parameterization scheme using lightning data as proxy for the presence of convection in an MM5 mesoscale model. The assimilation of lightning proved to have a positive impact on the representation of the precipitation field, also providing more realistic positioning of the precipitation maxima.

Following this example, various simulations of SOP1 case studies are expected to be performed based on WRF model. The WRF model (Skamarock et al., 2008) is a community mesoscale NWP model designed to be a flexible, state-of-the-art tool that is portable and computationally efficient on a wide variety of platforms. It is a fully compressible non-hydrostatic model with a terrain following a hydrostatic pressure vertical coordinate system and Arakawa C grid staggering. It is in the authors' plans to also investigate the ability of WRF model to predict the spatial and temporal distribution of lightning flashes based on the implemented scheme proposed by Barthe and Barth (2008), where the prediction of lightning flash rate is based on the fluxes of non-precipitating and precipitating ice.

4 Observations collected during the HyMeX SOP1

The following section presents an overview of observations collected by different PEACH instruments and demonstrates the rather comprehensive and unique data set of natural lightning flashes collected so far in Europe. The different examples shown here are not related to any other HyMeX SOP1 observations as the main goal of the paper focuses on the actual PEACH observations and their consistency. Several studies are already underway to relate the lightning activity and the electrical properties to microphysical and dynamical properties of the parent thunderclouds using observations from operational and research radars (e.g., Bousquet et al.,

2014), in situ airborne and ground-based probes and satellites, and numerical simulations.

4.1 SOP1 climatology

Figure 2 shows a comparison of the lightning activity as sensed by Météorage, the French component of the EUCLID network, over southeastern France for the period September-October-November (SON) 2012 and for the period 1997–2012. It is based on the number of days with at least one lightning flash recorded per day in a regular grid of 5 km × 5 km and cumulated over the period investigated. Only flashes identified as CG flashes by Météorage algorithms are considered here. A similar climatology, but for the period 1997–2011, was used to determine the most statistically electrically active area in the field domain to deploy and operate the lightning research sensors. Although further investigations on the climatologic properties of the lightning activity are underway, Fig. 2b shows the contribution of the 2012 records to the period 1997–2012. The year 2012 was rather weak in terms of lightning activity over the center of the SOP1 domain. The electrical activity was mainly located in the far northern part of CV domain and was more pronounced along the Riviera coastline and over the Ligurian Sea (Fig. 2b). About 0.3 % of the 5 km × 5 km pixels of the year 2012 contribute to more than 20 % of the 16-year climatology. Over the 500 km side domain plotted in Fig. 2a and b and for a period ranging from 5 September to 6 November, the total number of days with lightning activity in 2012 reached a value of 44 days, slightly below the average value for the 16 years of interest (Fig. 2c). Even if the lightning activity was less pronounced in 2012 over the CV domain, electrical properties of several convective systems were documented during SOP1 as shown in the following as well as in Ducrocq et al. (2014) and Bousquet et al. (2014). HyLMA also captured summer thunderstorms as it was already operated before the SOP1. During the deployment of the HyLMA network, and based on the experience gained during the Deep Convective Clouds and Chemistry (DC3) project, it was decided to enhance the actual coverage of the HyLMA network by deploying four of the 12 stations (Candillargues, Mont Aigoual, Mont Perier, Mirabel) away from the dense eight-station network. The redeployment to the west was also strongly recommended by the local Weather Office to document the growth of new electrical cells within V-shape storm complexes that usually occur in the southwestern zone of the field domain. Interestingly, this new configuration offered the possibility to record farther lightning activity in all directions.

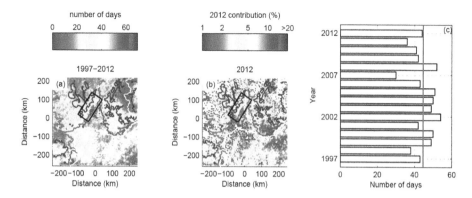

Figure 2. Cloud-to-ground lightning climatology in terms of number of days with at least one cloud-to-ground lightning flash recorded per day in a regular grid of 5 km × 5 km and cumulated over the period investigated as sensed by Météorage from 1997 to 2012 (**a**), contribution of the 2012 records expressed in % relative to the 1997–2012 number of days per 5 km × 5 km pixel (**b**) and number of days per year (**c**) for the period September–November over southeastern France. The red solid line plotted in (**c**) corresponds to the average value for the 1997–2012 period. Red and black red lines in (**a**) and (**b**) indicate 200 and 1000 m height, respectively. The Cévennes–Vivarais domain is also delimited by the black polygon.

4.2 Examples of concurrent PEACH observations

4.2.1 Flash level

A regular IC (24 September 2012 02:02:32 UT)

Figure 3 shows an example of a regular IC flash recorded by HyLMA during SOP1 Intensive Observation Period (IOP) IOP-06 on 24 September 2012. This flash was recorded within a mature convective cell. The lightning flash lasted for 800 ms. It was composed of 2510 VHF sources as reconstructed from at least seven HyLMA stations and $\chi^2 < 1$. For more information on the definition of the parameters associated to each LMA source the interested reader is referred to Thomas et al. (2004). The VHF sources were vertically distributed between 4 and 12 km (Fig. 3d). The IC flash was triggered at 8.5 km height (Fig. 3e). This IC flash exhibits a regular bi-level structure with long horizontal branches propagating at 6 and 11 km a.s.l. height (Fig. 3b and c). The lower branches show weaker VHF sources than the upper branches and spread over a larger altitude range (Fig. 3f). The high (low) altitude horizontal branches correspond to negative (positive) leaders propagating through positive (negative) charge regions. As expected the upper channels, i.e., negative leaders, propagated faster as evidenced from the actual distances traveled by the negative leaders compared to the ones traveled by the positive leaders during the same temporal gap. During the development of the flash, most of the breakdown events are detected by HyLMA at the edge of the discharges previously ionized and consequently tend to widen the lower and upper channels away from the upward channel. HyLMA partially mapped one fast process at 02:02:33.557 that lasted for 3.5 ms and propagated over 25 km from the lower to the upper part of the flash (see the black lines in Fig. 3). Finally, none of the OLLSs re-ported that specific IC flash, while other IC flashes have been recorded by the OLLSs.

A regular −CG (24 September 2012 01:43:17 UT)

Figure 4 shows a compilation of records for a multi-stroke −CG flash as recorded not only by HyLMA and the different OLLSs, but also as sampled at close range (25 km) by the VFRS instruments and one of the SLAs. The flash lasted for more than 1.1 s and was composed of nine connections to the ground as deduced from the VFRS data analysis (Fig. 4e and f). HyLMA reconstructed 1464 VHF sources derived from at least seven HyLMA stations. The VHF sources were all located below 5.5 km height (Fig. 4d), and their 3-D distribution indicates that a negative charge region was located south of the ground strokes at an average altitude of 4.5 km height (Fig. 4a–c). Note that for the present −CG flash, HyLMA did not map entirely the downward stepped leaders down to the ground (Fig. 4e and f).

The −CG flash was recorded by all OLLSs but ZEUS (Fig. 4g). ATDnet reported seven events, whereas EUCLID identified five strokes as negative ground connections and LINET categorized eight strokes as negative ground connections and one stroke as positive ground connection. Times of OLLS records obviously coincide with times of field record stroke measurements (in gray in Fig. 4e–g). The signal recorded by the SLA documented the changes induced by the successive ground connections and confirmed the negative polarity of the CG flash (Fig. 4f). The events recorded by the different OLLSs are mainly located close to each other except for one ATDnet stroke (Fig. 4a–c). Further investigations are underway to study both flash and stroke detection efficiencies and location accuracy of the OLLSs over the HyLMA domain using other coincident VFRS, SLA and HyLMA records.

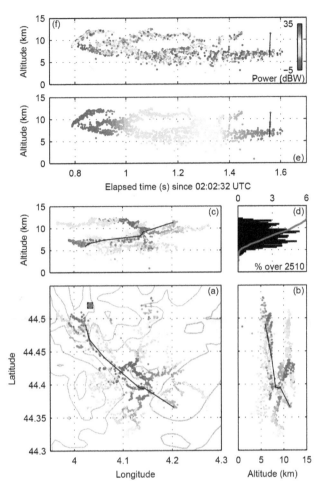

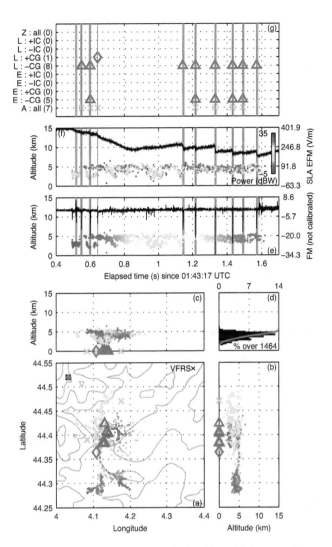

Figure 3. HyLMA records during a regular IC flash (24 September 2012, 02:02:32 UTC) with (**a**) ground projection of the lightning records with 200 m increment relief isolines; (**b**) latitude–altitude projection of the lightning records; (**c**) longitude–altitude projection of the lightning records; (**d**) 250 m increment histogram (bars) and cumulative distribution (red cure) of the VHF source altitude; (**e**) time–height series of VHF sources; and (**f**) amplitude–height series of VHF sources. The black lines join the successive VHF sources recorded during the *K* change event at 02:02:33.557 UTC.

Figure 4. Records during a −CG flash with multiple ground connections (24 September 2012, 01:43:17 UTC) with (**a**) ground projection of the lightning records; (**b**) latitude–altitude projection of the lightning records; (**c**) longitude–altitude projection of the lightning records; (**d**) histogram (bars) and cumulative distribution (red cure) of the VHF source altitude; (**e**) time–height series of VHF sources and record of the Uzès SLA; (**f**) amplitude–height series of VHF sources and record of the VFRS electric field observations; and (**g**) records of OLLSs per instrument and type of detected events available only for EUCLID and LINET. The orange bars correspond to ground strokes as identified from VFRS records. The VFRS location is also indicated in (**a**). Gray lines indicate times of all OLLS reports. Records from ATDnet, EUCLID, LINET and ZEUS are plotted with green crosses, blue symbols, red symbols and black stars, respectively.

The same CG flash was also documented with the 5 ms camera as shown in Fig. 5, where the images recorded at the time of the ground connections identified from VFRS records are compiled. Times of the successive (single) frames are indicated in orange in Fig. 4g. The two first frames in Fig. 5 show clearly two channels connecting to the ground. The other frames show scattered light accompanying the successive return strokes but with the channel itself masked by a nearby hill, except the frame at 01:43:18.490 where much weaker optical signal was recorded (Fig. 5). ATDnet, EUCLID and LINET detected this specific stroke (Fig. 4g) as well as the field record sensor (Fig. 4e), but the change induced by this stroke had little impact as detected with the SLA (Fig. 4f). The first channel to ground was recorded with-

out any question by the video camera but was not located by any OLLS. Interestingly, a flash located 42 km away from VFRS and north to the −CG flash triggered around the time of the first ground connection, so the radiation might have interfered with the signal radiated by the first ground connection. Additionally, the noisy field record signal recorded at

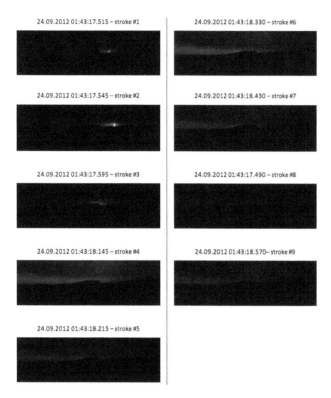

Figure 5. Enhanced VFRS 5 ms frames recorded during the nine ground strokes of the −CG flash presented in Fig. 4.

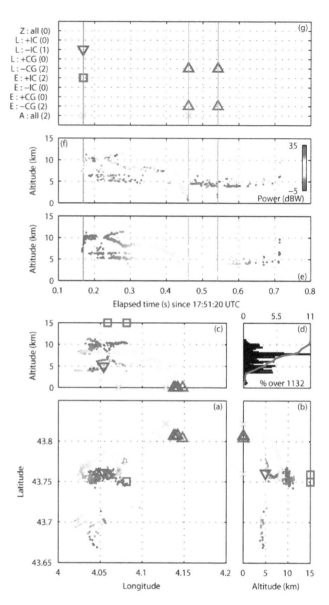

Figure 6. Concurrent lightning records during a bolt-from-the-blue flash recorded on 5 September 2012 at 17:51:20 UTC. See Fig. 4 for a description of each panel.

01:43:18.6 (Fig. 4e, elapsed time equal 1.6 s) emanated from the early stage of a 700 ms duration IC flash located 30 km north from the documented −CG flash.

Over the entire SOP1 campaign, several optical observations are available for other −CG flashes, +CG flashes and also IC flashes propagating along or below the cloud base. Even if the VFRS was mobile, it was often difficult to capture optical measurements either because of rain or presence of low-level clouds between the lightning flashes and the video camera. However, the recorded field record observations, with and without optical measurements, of the mobile instrumentations in conjunction with SLA records offer a rather unique ground truth to validate the OLLS records, quantify their detection efficiency and investigate in detail the flash processes that are recorded and located by the different OLLSs operated with short and long baselines.

Examples of unusual lightning flashes

The more HyLMA data are analyzed, the more we find lightning flashes that do not fit with either the bi-level structure of regular IC flashes or with the typical development of multistroke −CG flashes. In the following we present two examples of unusual lightning flashes. For instance, Fig. 6 presents the HyLMA and OLLS records for a specific type of flash called a bolt-from-the-blue (BFTB) type. In the present case, the flash (5 September 2012 17:51:20 UT) started like a reg-

ular IC flash with an ignition at 6 km height. The upper discharge split in two parts 50 ms after its ignition, one progressing continuously upward while the other went downward, propagating at a constant altitude of 8 km during 50 ms before descending and eventually connecting to the ground. The altitude–latitude panel (Fig. 6b) clearly shows several branches of negative stepped leaders approaching the ground while the flash propagates to the ground.

EUCLID and LINET reported the first ground connection and a second ground strike (Fig. 6e and f). Additionally, EUCLID and LINET reported IC events a few milliseconds after the first VHF source (Fig. 6g). The locations of the IC events given by EUCLID and LINET are consistent with the

HyLMA locations. LINET reported an IC event at an altitude of 5 km, just above the negative charge region. This example demonstrates the capability of operational systems like EUCLID and LINET to detect IC components and potentially IC flashes.

The ZEUS network did not locate any event during that specific flash. ATDnet recorded the first ground connection but also the VLF radiation in the early beginning of the flash with a rather accurate location (Fig. 6a–c). This example, among others, confirms the capability of sferic-detecting networks to locate some IC components, as Lagouvardos et al. (2009) already reported with ZEUS and LINET. The HyMeX SOP1 data offer a unique opportunity to not only study the CG and IC detection efficiencies as well as location accuracy, but also to investigate the discharge properties with a signal strong and well pronounced enough to be detected and located by long range VLF detection systems. Over a total of 124 flashes, 11 BFTB flashes were recorded during the entire life cycle of the isolated storm on 5 September 2012, with negative downward stepped leaders propagating from the upper positive charge region to the ground. Other BFTB flashes have been identified in the HyLMA data set analyzed so far, such as the ones observed during the IOP-06 case on 24 September 2012 (not shown).

Figure 7 presents an example of a complex flash recorded on 30 August 2012 (04:35:00 UTC) before the beginning of SOP1. The VHF radiations were recorded over more than 5 s and the lightning flash propagated from the northwest to the southeast over a large domain (> 120 km long; Fig. 7a–c). The temporal and spatial evolution of the successive discharges mapped by HyLMA reveals that the continuous VHF signal emanated from a single but extensive lightning flash. The flash mainly occurred on the eastern side of the HyLMA coverage area. Comparison with radar observations indicated that the flash propagated in a stratiform region (not shown). The spatial distribution of the VHF sources suggests the existence of multiple charge regions in the parent cloud at different altitudes (Fig. 7b and c). Another long-lasting flash occurred in the same area 4 (17) s before (after) the occurrence of the studied flash. Flashes of 2 to 3 s duration were also recorded between 04:00 and 05:00 UTC mostly in the northwestern part of the storm complex. Between 04:30 and 04:40 UTC, 44 flashes were recorded over the domain of interest; all but the one shown in Fig. 7 occurred in the northwestern electrical cell centered at 44.5° N and 5° E.

All OLLSs reported space and time consistent observations relative to HyLMA records. ATDnet reported four fixes, EUCLID 14 events including eight negative ground strokes and one positive ground stroke, LINET 14 events, all identified as ground strokes as no altitude information was available, and ZEUS seven fixes. A single flash identified by HyLMA is actually seen as multiple flashes by the OLLSs with the algorithms used to combine strokes/fixes into flashes. This unusual flash example demonstrates the relevance and the usefulness of VHF mapping to characterize

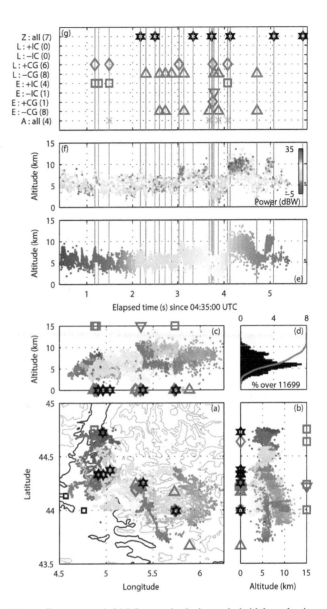

Figure 7. LMA and OLLS records during a hybrid long-lasting flash. See Fig. 4 for a description of each panel. The relief is plotted with 500 m isolines. The black isoline corresponds to 200 m height.

the full 3-D spatial extension of the lightning flashes. Additionally, some of the events detected by one OLLS are also detected by one or more other OLLSs, while sometimes an event is reported by a single OLLS only. This was also observed during the analysis of the lightning data for the 6–8 September 2010 storm but not discussed in Lagouvardos et al. (2013). Such discrepancies are explained by the differences between the four OLLSs in terms of technology, range and amplitude sensibility, detection efficiency and location algorithms. For the studied flash, coincident OLLS strokes are observed with a time difference from 60 to 130 μs between long-range and short-range OLLSs and around 20 μs between EUCLID and LINET.

Concurrent VHF and acoustic measurements

Acoustic and infrasonic measurements were performed during HyMeX SOP1 as detailed in Sect. 3.2.3. Figure 8 presents an example of concurrent records during 2.5 min of lightning activity sensed on 24 September 2012. During that period, HyLMA detected seven lightning flashes (with one composed of a few VHF sources) in the studied area (Fig. 8a and e), all inducing a moderate to significant change on the SLA signal (Fig. 8g). ATDnet sensed all flashes except the one composed of a few VHF sources at $T = 48$ s (Fig. 8g). EUCLID, LINET and ZEUS recorded all but two flashes including the one composed of a few VHF sources; the second flash was not the same for these three OLLSs. ZEUS erroneously located additional flashes in the domain of interest. Among the seven flashes, three were connected to the ground with a negative polarity (Fig. 8g). The lightning activity was located about 20 km away from the acoustic sensors marked with a red diamond in Fig. 8a. The time evolution of the pressure difference (Fig. 8e) traces two acoustic events of duration greater than 20 s. The first event, between $T = 40$ s and $T = 70$ s is related to the first IC flash recorded during the first seconds of the studied period. The second acoustic event, starting at $T = 105$ s, comes from the two flashes (one $-$CG and one IC) recorded between $T = 60$ s and $T = 70$ s. The propagation of sound waves in the atmosphere and the properties of the atmosphere along the acoustic path to the acoustic sensors are at the origin of the delay between the recording of the electromagnetic signal and the recording of the acoustic signal. For the first acoustic event, the acoustic spectrogram (Fig. 8f) reveals a series of three acoustic bursts while for the second acoustic event, the spectrogram shows a less powerful signal. A signal of 0.2 Pa (absolute value) received by the sensors 20 km away from the storm is in the amplitude range of acoustical signals usually recorded. Based on the unique data set collected during the SOP1, several studies have been performed to relate the acoustic signal and its spectral and temporal properties to the original lightning flash type and properties.

4.2.2 Examples of SOP1 daily lightning activity recorded by HyLMA

The previous sections showed a series of concurrent records at the flash scale. Here we discuss some storms recorded during the SOP1. Although lightning activity recorded during the June–August period is not discussed here, it is worth mentioning that different types of storms were fully recorded during the entire HyLMA operation. As an example, Fig. 9 shows daily lightning maps as produced only from HyLMA data with, for each considered day, the 10 min VHF source rate reconstructed from at least seven LMA stations over the HyLMA coverage area in panel a, the geographical distribution of the lightning activity (the grayscale is time related) with an overlay of the 1 h VHF source density (per

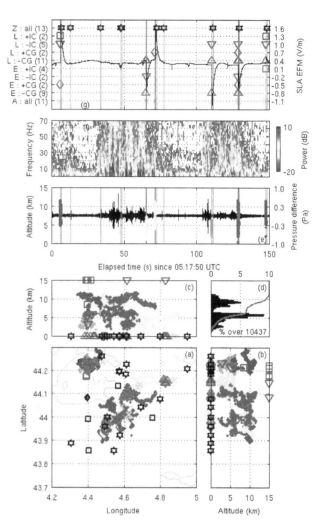

Figure 8. Coincident observations recorded between 05:17:50 and 05:20:20 UTC on 24 September 2012 with **(a)** ground projection of the lightning records; **(b)** latitude–altitude projection of the lightning records; **(c)** longitude–altitude projection of the lightning records; **(d)** 250 m increment histogram (bars) and cumulative distribution (red cure) of the VHF source altitude; **(e)** time–height series of VHF sources and pressure difference measured at the MPA location; **(f)** time series of the acoustic spectrum as recorded at MPA location; and **(g)** records of OLLSs per instrument with the time series of the Uzès SLA record.

$0.025° \times 0.025°$) at one specific hour in panel b, and the vertical distribution of the VHF sources (per $0.025° \times 200$ m) computed during the hour indicated at the top of the figure in panel c. As already mentioned, different types of convective systems were recorded during the operation of HyLMA, ranging from gentle isolated thunderstorms to organized and highly electrical convective lines between June 2012 and November 2012.

Figure 9A shows the lightning activity recorded during IOP-01 (11 September 2012) associated with scattered deep convection developing in early afternoon (Fig. 9A.a) over southeastern Massif Central and due to a convergence between a slow southeasterly flow from the Mediterranean Sea and a westerly flow from the Atlantic. The convection remained isolated and mainly confined to mountainous areas, with some cells reaching the foothills in late afternoon due to the westerly mid-level flow (Fig. 9A.b). The French F20 research aircraft, along with the airborne 95 GHz Doppler cloud radar RASTA (RAdar SysTem Airborne) and in situ microphysics probes, sampled the anvils of the closest convective cells to the HyLMA stations. The rainfall accumulation ranged from 5 to 10 mm in 24 h and reached local levels of up to 30–40 mm in Ardèche. This example shows typical observations collected with HyLMA during scattered convection over the domain of interest, definitively demonstrating that the records of HyLMA as well as the records of OLLSs offer the possibility of a radar-like tracking of storm motions.

Figure 9B shows the HyLMA records during IOP-06 (24 September 2012). An intense and fast-moving convective line crossed the CV domain during the early morning, Liguria–Tuscany by mid-day and northeastern Italy in the evening, with an amount of rainfall observed of 100 mm in 24 h over southeastern France, rainfall intensity up to 50–60 mm h^{-1} and wind gusts up to 90–100 km h^{-1} locally. The storm activity started in the evening of 23 September on the west side of the HyLMA network and moved to the east with successive electrical cells developing and merging. Figure 9B.b and c show one of the highest density of VHF sources recorded during the entire period of HyLMA operation. Between 02:00 and 03:00 UTC, the lightning activity was more or less distributed along a north–south direction but then extended further north to the HyLMA network (Fig. 9B.b). Focusing on the electrical cells located in the vicinity of the LMA network, the lightning activity was located east of strong updrafts retrieved from the radar data (see Fig. 8 in Bousquet et al., 2014), with the deepest electrified convective cell reaching up to 13 km height. Many different PEACH instruments documented the lightning activity of this storm as shown in Figs. 3, 4 and 8. The VFRS was operated from the Aubenas airfield (44.538° N, 4.371° E) from the early hours of the storm activity to mid-morning. Some storm cells were also documented with the airborne RASTA radar and in situ microphysics probes on board the F20 and by different precipitation research radars located in the northern part of the HyLMA coverage area.

Figure 9C shows the total lightning activity sensed during IOP-07a (26 September 2012). The first convective system appears early in the morning over the HyLMA because of a warm, unstable and convergent air mass that merges with a frontal system progressing eastwards during the afternoon. This event brings more than 100 mm in 24 h over the CV region. Additionally, the city of Nice on the Riviera Coast was flooded in the evening (Fig. 9C.b). The VFRS operated from Valence (44.992° N 4.887° E) during the first part of the day and then moved to Mont Ventoux (44.171° N, 5.202° E).

During the morning observations, most CG flashes recorded with VFRS instruments in the northern part of the convective complex were of positive polarity, while the CG flashes in the afternoon were mostly negative.

Figure 9D shows the HyLMA records for 29 September 2012 (IOP-08). This system moved from Spain where heavy precipitation was recorded on the northeasterly flank of Spain with casualties and significant damages. Figure 9D.b shows an extensive area of coverage by the HyLMA in its southeastern sector with more pronounced altitude errors for very distant flashes. The case is interesting as it moved from sea to land (Fig. 9D.b); it allows the investigation of contrasting lightning properties over sea and over land as well as the documentation of the transition from sea to land. VFRS observations were collected for lightning flashes along the Riviera.

Figure 9E shows the lightning activity of IOP-13 (14 October 2012) where Nice airport was closed at the end of the day because of strong vertical shear. A tornado (EF1) was observed in the vicinity of Marseille between 14:00 and 15:00 UTC. The analysis of the lightning activity of the tornado cloud revealed the occurrence of a convective surge with a sudden increase of the flash rate and an upward shift of the flash triggering altitude (not shown). Analyses combining HyLMA, OLLSs and operational radar records are underway to evaluate the benefit of lightning detection in terms of information precursors related to this tornado. Additionally, the French F20 aircraft sampled some electrified clouds and later (17:00–20:00 UTC) performed a survey of precipitating systems over Provence/Côte d'Azur (French Riviera), which offers the possibility to study in situ microphysics, vertical structure of the clouds and lightning activity.

Finally, Fig. 9F shows the observed lightning activity during IOP-16a (26 October 2012). A system first affected the Hérault and Gard departments in the morning but a second more intense system developed in the southeast of France in the afternoon with two casualties in Toulon. Rain accumulation reached up to 170 mm in 24 h in the CV domain. The F20 aircraft flew between 06:00 and 09:30 UTC in the complex located at 43° N, 4° E (Fig. 9F.b). A second F20 flight sampled the electrically active storms shown in Fig. 9F.b (43.2° N, 6° E; 43.2° N, 3° E). VFRS observations were performed at the end of the day about 50 km east of the HyLMA network for a series of mainly −CG flashes. Between 20:30 and 20:40 UTC the lightning activity sensed in the vicinity of the MBA/MPA network was rather weak (i.e., 24 flashes in 10 min), so one-to-one correlations between RF HyLMA and EUCLID records and non-noisy acoustics signals from the same flashes are currently being studied (not shown).

5 Prospects

The present article summarizes only a small number of events observed with different PEACH instruments during

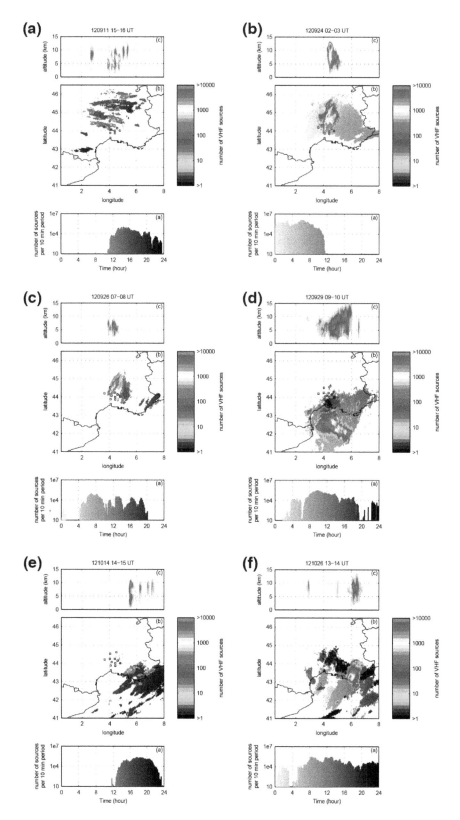

Figure 9. Total lightning activity recorded at different dates with HyLMA. (**a**) HyLMA VHF source rate per 10 min period (plotted in decimal logarithmic scale); (**b**) ground projection of the HyLMA sources during 24 h (in gray, from 00:00 to 23:59 UTC) and density of HyLMA VHF sources during 1 h computed per $0.025° \times 0.025°$ grid (in color); (**c**) vertical distribution of the HyLMA VHF sources for the same 1 h period (and indicated at the top of the panel) per $0.025° \times 200$ m grid.

HyMeX SOP1. This rather unique and comprehensive lightning data set collected during the SOP1 will serve to investigate the properties of individual lightning flashes but also to probe objectively, for the first time, the performances of European OLLSs in southeastern France and close to the Mediterranean Sea. This task will help to refine our current knowledge on what European OLLSs actually record and more specifically which intra-cloud processes are detected and located. The investigation should eventually provide new insights on the potential of IC detection from European OLLSs for operational storm tracking and monitoring over the entire Mediterranean basin.

Several analyses are already underway to investigate the properties of the lightning activity from the flash scale to the regional scale in relation to cloud and atmospheric properties as derived from satellite imagery, operational/research ground-based and airborne radars, rain gauges and in situ microphysical probes. The analyses focus not only on HyMeX SOP1 priority cases (Ducrocq et al., 2014) but also on non-SOP1 events as HyLMA data cover June 2012 to the end of November 2012. The analysis will eventually provide key lightning-related indexes to describe the electrical nature of thunderstorms in southeastern France and which will be used in multi-disciplinary studies carried out within HyMeX. The combination of HyLMA and OLLS records will provide a set of basic products – e.g., flash rate, flash type, flash properties and flash density – to populate the HyMeX database.

The HyMeX case studies are not only observationally oriented but are also intended to provide material for verification and validation of kilometer-scale electrified cloud simulations (e.g., Pinty et al., 2013). Indeed, successful simulations are already performed and comparisons of simulated and observed parameters – e.g., vertical distribution of the charge regions, flash location, flash rate and flash extension – are already showing promising results. The HyLMA data should then help to identify objectively which non-inductive charging process treatment ("Takahashi" versus "Saunders") leads to the best simulation results.

An objective debriefing of SOP1 preparation, operation and data analysis will be performed in the near future to identify the successes and the failures. This will help us to refine the preparation of a dedicated atmospheric electricity field campaign in early autumn 2016 over Corsica because a permanent LMA was established there in May 2014 for a minimum of 5 years. Another region of interest is the eastern Mediterranean Sea during autumn, when electrical activity takes place over the sea but ceases when the thunderclouds are landing.

Finally, the different activities performed around the PEACH project have already helped us gain expertise not only for field deployment and operations but also in terms of data analysis methodologies, realistic lightning and cloud simulations and application of lightning detection for very short-range forecasts in preparation for the EUMETSAT Meteosat Third Generation Lightning Imager (launch scheduled early 2019).

Acknowledgements. This project was sponsored by grants MISTRALS/HyMeX and ANR-11-BS56-0005 IODA-MED. LEFE-IDAO, Université de Toulouse and the GOES-R visiting program also supported the PEACH project during its preparation and the field campaign. The Greek contribution to PEACH objectives is partially funded by the TALOS project, in the frame of "ARISTEIA II", by the Greek General Secretariat for Research and Technology. We are grateful to R. Blakeslee and NASA for lending the MSFC LMA during the SOP1. We would like to thank the team of the local Météo-France weather office in Nîmes for its strong support during the site survey, deployment, operation and dismantlement of the instruments and for letting us deploy four of the HyLMA stations on Météo-France land. We also thank Mr. and Ms. Imbert (Cadignac HyLMA site), Mr. Rey and the Méjannes-le-Clap City Council (Méjannes-le-Clap HyLMA site), Mr. Comte (Vic-le-Fesq HyLMA site), Mr. and Ms. Bazalgette (Mont Aigoual HyLMA site), Mr. Vincent (Mont Perier HyLMA site), Mr. Chaussedent (Mirabel HyLMA site), Mr. Fourdrigniez (CCI Alès Deaux airfield HyLMA site) and Mr. Garrouste (CNRM-GAME, responsible for the Candillargues HyMeX supersite; Candillargues HyLMA site) for hosting a HyLMA station. We also thank Mr. Reboulet (Mayor of La Bruguière) for allowing the deployment of one SLA and the MBA/MPA package on his property. We also thank Mr. Cerpedes (La Grande-Combe Technical Manager) for letting us deploy the second SLA on La Grande-Combe airfield. We are also grateful to the different weather forecasters and the HyMeX operation direction for the support to the VFRS. We thank Georg Pistotnik from the European Severe Storm Laboratory (ESSL) for providing additional special forecasts in preparation and during several VFRS observation trips. We thank Patrice Blanchet (ONERA) and Philippe Lalande (ONERA) for lending us their fast video camera that was operated at some of the VFRS sites. We also thank Brice Boudevilain (LTHE) and Olivier Bousquet (Météo-France) for providing contacts for the deployment of the four most remote HyLMA stations. Finally we would like to thank D. Poelman and the second anonymous reviewer for their careful reading of the manuscript and their constructive suggestions.

Edited by: M. Nicolls

References

Adamo, C.: On the use of lightning measurements for the microphysical analysis and characterization of intense precipitation events over the Mediterranean area, PhD dissertation, Univ. of Ferrara, Ferrara, Italy, 2004.

Altaratz, O., Reisin, T., and Levin, Z.: Simulation of the Electrification of Winter Thunderclouds using the 3-dimensional RAMS Model: single cloud simulations, J. Geophys. Res, 110, 1–12, D20205, 2005.

Arechiga, R. O., Johnson, J. B., Edens, H. E., Thomas, R. J., and Rison, W.: Acoustic localization of triggered lightning, J. Geophys. Res., 116, D09103, doi:10.1029/2010JD015248, 2011.

Barthe, C. and Barth, M. C.: Evaluation of a new lightning-produced NO_x parameterization for cloud resolving models and its associated uncertainties, Atmos. Chem. Phys., 8, 4691–4710, doi:10.5194/acp-8-4691-2008, 2008.

Barthe, C. and Pinty, J.-P.: Simulation of electrified storms with comparison of the charge structure and lightning efficiency, J. Geophys. Res., 112, D19204, doi:10.1029/2006JD008241, 2007a.

Barthe, C. and Pinty, J.-P.: Simulation of a supercellular storm using a three-dimensional mesoscale model with an explicit lightning flash scheme, J. Geophys. Res., 112, D06210, doi:10.1029/2006JD007484, 2007b.

Barthe, C., Molinié, G., and Pinty, J.-P.: Description and first results of an explicit electrical scheme in a 3D cloud resolving model, Atmos. Res., 76, 95–113, 2005.

Barthe, C., Pinty, J.-P., and Mari, C.: Lightning-produced NO_x in an explicit electrical scheme tested in a Stratosphere-Troposphere Experiment: Radiation, Aerosols, and Ozone case study, J. Geophys. Res., 112, D04302, doi:10.1029/2006JD007402, 2007.

Barthe, C., Chong, M., Pinty, J.-P., Bovalo, C., and Escobar, J.: CELLS v1.0: updated and parallelized version of an electrical scheme to simulate multiple electrified clouds and flashes over large domains, Geosci. Model Dev., 5, 167–184, doi:10.5194/gmd-5-167-2012, 2012.

Beard, K. V.: Terminal velocity and shape of cloud and precipitation drops aloft, J. Atmos. Sci., 33, 851–864, 1976.

Betz, H.-D., Schmidt, K., and Oettinger, W. P.: LINET – An International VLF/LF Lightning Detection Network in Europe, in: Lightning: Principles, Instruments and Applications, edited by: Betz, H.-D., Schumann, U., and Laroche, P., Ch. 5, Dordrecht (NL), Springer, 2008.

Betz, H.-D., Schmidt, K., Laroche, P., Blanchet, P., Oettinger, W. P., Defer, E., Dziewit, Z., and Konarski, J.: LINET – an international lightning detection network in Europe, Atmos. Res., 91, 564–573, 2009.

Blanc, E., Farges, T., Le Pichon, A., and Heinrich, P.: Ten year observations of gravity waves from thunderstorms in western Africa, J. Geophys. Res.-Atmos., 119, 6409–6418, doi:10.1002/2013JD020499, 2014.

Bousquet, O., Berne, A., Delanoe, J., Dufournet, Y., Gourley, J. J., Van-Baelen, J., Augros, C., Besson, L., Boudevillain, B., Caumont, O., Defer, E., Grazioli, J., Jorgensen, D. J., Kirstetter, P.-E., Ribaud, J.-F., Beck, J., Delrieu, G., Ducrocq, V., Scipion, D., Schwarzenboeck, A., and Zwiebel, J.: Multiple-Frequency Radar Observations Collected In Southern France During HyMeX SOP-1, B. Am. Meteorol. Soc., online first, doi:10.1175/BAMS-D-13-00076.1, 2014.

Chauzy, S. and Soula, S.: General interpretation of surface electric field variations between lightning flashes, J. Geophys. Res., 92, 5676–5684, 1987.

Christian, H. J., Blakeslee, R. J., Boccippio, D. J., Boeck, W. L., Buechler, D. E., Driscoll, K. T., Goodman, S. J., Hall, J. M., Koshak, W. J., Mach, D. M., and Stewart, M. F.: Global frequency and distribution of lightning as observed from space by the Optical Transient Detector, J. Geophys. Res., 108, 4005, doi:10.1029/2002JD002347, 2003.

Coquillat, S. and Chauzy, S.: Computed conditions of corona emission from raindrops, Corona emission from raindrops in strong electric fields as a possible discharge initiation: comparison between horizontal and vertical field configurations, J. Geophys. Res., 99, 16897–16905, 1994.

Coquillat, S., Combal, B., and Chauzy, S.: Corona emission from raindrops in strong electric fields as a possible discharge initiation: comparison between horizontal and vertical field configurations, J. Geophys. Res., 108, 4205, doi:10.1029/2002JD002714, 2003.

Crabb, J. A. and Latham, J.: Corona from colliding drops as a possible mechanism for the triggering of lightning, Q. J. Roy. Meteor. Soc., 100, 191–202, 1974.

Cummins, K., Murphy, M., Bardo, E., Hiscox, W., Pyle, R., and Pifer, A.: A Combined TOA-MDF Technology Upgrade of the U.S. National Lightning Detection Network, J. Geophys. Res., 103, 9035–9044, 1998.

Defer, E. and Laroche, P.: Observation and Interpretation of Lightning Flashes with Electromagnetic Lightning Mapper, in: Lightning: Principles, Instruments and Applications, edited by: Betz, H.-D., Schumann, U., and Laroche, P., Ch. 5, Dordrecht (NL), Springer, 2009.

Defer, E., Blanchet, P., Théry, C., Laroche, P., Dye, J., Venticinque, M., and Cummins, K.: Lightning activity for the July 10, 1996, storm during the Stratosphere-Troposphere Experiment: Radiation, Aerosol, and Ozone-A (STERAO-A) experiment, J. Geophys. Res., 106, 10151–10172, 2001.

Defer, E., Lagouvardos, K., and Kotroni, V.: Lightning activity in Europe as sensed by long range NOA-ZEUS and UK Met Office ATD VLF lightning systems and NASA TRMM-LIS sensor, Geophys. Res. Abstr., EGU05-A-03026, EGU General Assembly 2005, Vienna, Austria, 2005.

Ducrocq, V., Braud, I., Davolio, S., Ferretti, R., Flamant, C., Jansa, A., Kalthoff, N., Richard, E., Taupier-Letage, I., Ayral, P.-A., Belamari, S., Berne, A., Borga, M., Boudevillain, B., Bock, O., Boichard, J.-L., Bouin, M.-N., Bousquet, O., Bouvier, C., Chiggiato, J., Cimini, D., Corsmeier, U., Coppola, L., Cocquerez, P., Defer, E., Delanoë, J., Di Girolamo, P., Doerenbecher, A., Drobinski, P., Dufournet, Y., Fourrié, N., Gourley, J. J., Labatut, L., Lambert, D., Le Coz, J., Marzano, F. S., Molinié, G., Montani, A., Nord, G., Nuret, M., Ramage, K., Rison, W., Roussot, O., Said, F., Schwarzenboeck, A., Testor, P., Van Baelen, J., Vincendon, B., Aran, M., and Tamayo, J.: HyMeX-SOP11: The Field Campaign Dedicated to Heavy Precipitation and Flash Flooding in the Northwestern Mediterranean, B. Am. Meteorol. Soc., 95, 1083–1100, doi:10.1175/BAMS-D-12-00244.1, 2014.

Duffourg, F. and Ducrocq, V.: Origin of the moisture feeding the Heavy Precipitating Systems over Southeastern France, Nat. Hazards Earth Syst. Sci., 11, 1163–1178, doi:10.5194/nhess-11-1163-2011, 2011.

Farges, T. and Blanc, E.: Characteristics of infrasound from lightning and sprites near thunderstorm areas, J. Geophys. Res., 115, A00E31, doi:10.1029/2009JA014700, 2010.

Füllekrug, M., Kolmasova, I., Santolik, O., Farges, T., Bor, J., Bennett, A., Parrot, M., Rison, W., Zanotti, F., Arnone, E., Mezentsev, A., Lan, R., Uhlir, L., Harrison, G., Soula, S., van der Velde, O., Pinçon, J.-L., Helling, C., and Diver, D.: Electron Acceleration Above Thunderclouds, Environ. Res. Lett., 8, 035027, doi:10.1088/1748-9326/8/3/035027, 2013.

Funatsu, B., Claud, C., and Chaboureau, J.-P.: Comparison between the Large-Scale Environments of Moderate and Intense Precipitating Systems in the Mediterranean Region, Mon. Weather Rev., 137, 3933–3959, 2009.

Gaffard, C., Nash, J., Atkinson, N., Bennett, A., Callaghan, G., Hibbett, E., Taylor, P., Turp, M., and Schulz, W.: Observing lightning around the globe from the surface, in: the Preprints, 20th International Lightning Detection Conference, Tucson, Arizona, 21–23, 2008.

Gallin, L.-J.: Caractérisation acoustique des éclairs d'orage, PhD dissertation, Université Pierre et Marie Curie, Paris, France, 2014.

Goodman, S. J., Buechler, D. E., Wright, P. D., and Rust, W. D.: Lightning and precipitation history of a microburst-producing storm, Geophys. Res. Lett., 15, 1185–1188, 1988.

Gurevich, A. V., Milikh, G. M., and Roussel-Dupre, G. M.: Runaway electron mechanism of air breakdown and preconditioning during a thunderstorm, Phys. Lett. A, 165, 463–468, 1992.

Helsdon Jr., J. and Farley, R.: A Numerical Modeling Study of a Montana Thunderstorm: 2. Model Results Versus Observations Involving Electrical Aspects, J. Geophys. Res., 92, 5661–5675, 1987.

Helsdon Jr., J. H., Gattaleeradapan, S., Farley, R. D., and Waits, C. C.: An examination of the convective charging hypothesis: Charge structure, electric fields, and Maxwell currents, J. Geophys. Res., 107, 4630, doi:10.1029/2001JD001495, 2002.

Holt, M. A., Hardaker, P. J., and McLelland, G. P.: A lightning climatology for Europe and the UK, 1990–99, Weather, 56, 290–296, 2001.

Jacobson, A. R., Knox, S. O., Franzand, R., and Enemark, D. C.: FORTE observations of lightning radio-frequency signatures: Capabilities and basic results, Radio Sci., 34, 337–354, 1999.

Kohn, M., Galanti, E., Price, C., Lagouvardos, K., and Kotroni, V.: Now-Casting Thunderstorms in the Mediterranean Region using Lightning Data, Atmos. Res., 100, 489–502, 2011.

Kotroni, V. and Lagouvardos, K.: Lightning occurrence in relation with elevation, terrain slope and vegetation cover in the Mediterranean, J. Geophys. Res., 113, D21118, doi:10.1029/2008JD010605, 2008.

Krehbiel, P. R., Thomas, R. J., Rison, W., Hamlin, T., Harlin, J., and Davis, M.: GPS-based mapping system reveals lightning inside storms, EOS, 81, 21–25, 2000.

Kummerow, C., Barnes, W., Kozu, T., Shiue, J., and Simpson, J.: The tropical rainfall measuring mission (TRMM) sensor package, J. Atmos. Ocean. Tech., 15, 809–817, 1998.

Lagouvardos, K., Kotroni, V., Betz, H.-D., and Schmidt, K.: A comparison of lightning data provided by ZEUS and LINET networks over Western Europe, Nat. Hazards Earth Syst. Sci., 9, 1713–1717, doi:10.5194/nhess-9-1713-2009, 2009.

Lagouvardos, K., Kotroni, V., Defer, E., and Bousquet, O.: Study of a heavy precipitation event over southern France, in the frame of HYMEX project: Observational analysis and model results using assimilation of lightning, Atmos. Res., 134, 45–55, 2013.

Light, T. E., Suszcynsky, D. M., and Jacobson, A. R.: Coincident radio frequency and optical emissions from lightning, observed with the FORTE satellite, J. Geophys. Res., 106, 28223–28231, doi:10.1029/2001JD000727, 2001.

MacGorman, D. R. and Rust, W. D.: The electrical nature of storms, Oxford University Press, New York, 422 pp., 1998.

MacGorman, D. R., Few, A. A., and Teer, T. L.: Layered lightning activity, J. Geophys. Res., 86, 9900–9910, doi:10.1029/JC086iC10p09900, 1981.

MacGorman, D. R., Burgess, D. W., Mazur, V., Rust, W. D., Taylor, W. L., and Johnson, B. C.: Lightning rates relative to tornadic storm evolution on 22 May 1981, J. Atmos. Sci., 46, 221–250, 1989.

Mäkelä, A., Tuomi, T. J., and Haapalainen, J.: A decade of high-latitude lightning location: Effects of the evolving location network in Finland, J. Geophys. Res., 115, D21124, doi:10.1029/2009JD012183, 2010.

Mansell, E. R., MacGorman, D. R., Ziegler, C. L., and Straka, J. M.: Simulated three-dimensional branched lightning in a numerical thunderstorm model, J. Geophys. Res., 107, 4075, doi:10.1029/2000JD000244, 2002.

Marshall, T. C., Stolzenburg, M., Maggio, C. R., Coleman, L. M., Krehbiel, P. R., Hamlin, T., Thomas, R. J., and Rison, W.: Observed electric fields associated with lightning initiation, Geophys. Res. Lett., 32, L03813, doi:10.1029/2004GL021802, 2005.

Molinié, G., Pinty, J.-P., and Roux, F.: Some explicit microphysical and electrical aspects of a Cloud Resolving Model: Description and thunderstorm case study, C. R. Physique, 3, 1–20, 2002.

Montanyà, J., Soula, S., and Pineda, N.: A study of the total lightning activity in two hailstorms, J. Geophys. Res., 112, D13118, doi:10.1029/2006JD007203, 2007.

Montanyà, J., Soula, S., Murphy, M., March, V., Aranguren, D., Solà, G., and Romero, D.: Estimation of charge neutralized by negative cloud-to-ground flashes in Catalonia thunderstorms, J. Electrostat., 67, 513–517, 2009.

Orville, R. E., Huffines, G. R., Burrows, W. R., and Cummins, K. L.: The North American Lightning Detection Network (NALDN) – Analysis of Flash Data: 2001–09, Mon. Weather Rev., 139, 1305–1322, 2011.

Pinty, J.-P., Barthe, C., Defer, E., Richard, E., and Chong, M.: Explicit simulation of electrified clouds: from idealized to real case studies, Atmos. Res., 123, 82–92, 2013.

Poeppel, K.: A 3D Lightning parameterization with branching and charge induction, Master's thesis, S. D. Sch. of Mines and Technol., Rapid City, 2005.

Price, C., Yair, Y., Mugnai, A., Lagouvardos, K., Llasat, M. C., Michaelides, S., Dayan, U., Dietrich, S., Di Paola, F., Galanti, E., Garrote, L., Harats, N., Katsanos, D., Kohn, M., Kotroni, V., Llasat-Botija, M., Lynn, B., Mediero, L., Morin, E., Nicolaides, K., Rozalis, S., Savvidou, K., and Ziv, B.: Using lightning data to better understand and predict flash floods in the Mediterranean, Surv. Geophys., 32, 733–751, 2011.

Proctor, D. E.: VHF radio pictures of cloud flashes, J. Geophys. Res., 86, 4041–4071, 1981.

Rison, W., Thomas, R. J., Krehbiel, P. R., Hamlin, T., and Harlin, J.: A GPS-based Three-Dimensional Lightning Mapping System: Initial Observations in Central New Mexico, Geophys. Res. Lett., 26, 3573–3576, 1999.

Rust, W. D., MacGorman, D. R., Bruning, E. C., Weiss, S. A., Krehbiel, P. R., Thomas, R. J., Rison, W., Hamlin, T., and Harlin, J.: Inverted-polarity electrical structures in thunderstorms in the Severe Thunderstorm Electrification and Precipitation Study (STEPS), Atmos. Res., 76, 247–271, 2005.

Saunders, C. P. R., Keith, W. D., and Mitzeva, P. P.: The effect of liquid water on thunderstorm charging, J. Geophys. Res., 96, 11007–11017, 1991.

Schroeder, V., Baker, M. B., and Latham, J.: A model study of corona emission from hydrometeors, Q. J. Roy. Meteor. Soc., 125, 1681–1693, 1999.

Schulz, W. and Saba, M. M. F.: First Results of Correlated Lightning Video Images and Electric Field Measurements in Austria, X International Symposium on Lightning Protection (SIPDA), Curitiba, Brazil, November, 2009.

Schulz, W., Lackenbauer, B., Pichler, H., and Diendorfer, G.: LLS Data and Correlated Continuous E-Field Measurements, VIII International Symposium on Lightning Protection (SIPDA), Sao Paulo, Brazil, 2005.

Schulz, W., Poelman, D., Pedeboy, S., Vergeiner, C., Pichler, H., Diendorfer, G., and Pack, S.: Performance validation of the European Lightning Location System EUCLID, International Colloquium on Lightning and Power Systems (CIGRE), Lyon, France, 2014.

Schultz, C., Petersen, W. A., and Carey, L. D.: Lightning and Severe Weather: A Comparison between Total and Cloud-to-Ground Lightning Trends, Weather Forecast., 26, 744–755, doi:10.1175/WAF-D-10-05026.1, 2011.

Shao, X.-M. and Krehbiel, P. R.: The spatial and temporal development of intracloud lightning, J. Geophys. Res., 101, 26641–26668, 1996.

Skamarock, W. C., Klemp, J. B., Dudhia, J., Gill, D. O., Barker, D. M., Duda, M. G., Huang, X.-Y., Wang, W., and Powers, J. G.: A description of the Advanced Research WRF version 3, NCAR Tech. Note NCAR/TN-475 + STR, 113 pp., 2008.

Smith D. A., Eack, K. B., Harlin, J., Heavner, M. J., Jacobson, A. R., Massey, R. S., Shao, X. M., and Wiens, K. C.: The Los Alamos Sferic Array: A research tool for lightning investigations, J. Geophys. Res., 107, 4183, doi:10.1029/2001JD000502, 2002.

Stephens, G. L., Vane, D. G., Boain, R. J., Mace, G. G., Sassen, K., Wang, Z., Illingworth, A. J., O'connor, E. J., Rossow, W. B., Durden, S. L., Miller, S. D., Austin, R. T., Benedetti, A., Mitrescu, C., and The Cloudsat Science Team: The Cloudsat mission and the A-Train, B. Am. Meteorol. Soc., 83, 1771–1790, doi:10.1175/BAMS-83-12-1771, 2002.

Stolzenburg, M., Rust, W. D., and Marshall, T. C.: Electrical structure in thunderstorm convective regions. 3. Synthesis, J. Geophys. Res., D103, 14097–14108, 1998.

Soula, S. and Georgis, J. F.: Surface electrical field evolution below the stratiform region of MCS storms, Atmos. Res., 132–133, 264–277, 2013.

Soula, S., Chauzy, S., Chong, M., Coquillat, S., Georgis, J. F., Seity, Y., and Tabary, P.: Surface precipitation current produced by convective rains during MAP, J. Geophys. Res., 108, 4395, doi:10.1029/2001JD001588, 2003.

Standler, R. B. and Winn, W. P.: Effects of coronae on electric fields beneath thunderstorms, Q. J. Roy. Meteor. Soc., 105, 285–302, 1979.

Takahashi, T.: Riming electrification as a charge generation mechanism in thunderstrom, J. Atmos. Sci., 35, 1536–1548, 1978.

Thomas, R. J., Krehbiel, P.R., Rison, W., Hunyady, S.J., Winn, W.P., Hamlin, T., and Harlin, J.: Accuracy of the Lightning Mapping Array, J. Geophys. Res., 109, D14207, doi:10.1029/2004JD004549, 2004.

Williams, E., Boldi, B., Matlin, A., Weber, M., Hodanish, S., Sharp, D., Goodman, S., Raghavan, R., and Buechler, D.: The behavior of total lightning activity in severe Florida thunderstorms, Atmos. Res., 51, 245–265, 1999.

Yair, Y., Lynn, B., Price, C., Kotroni, V., Lagouvardos, K., Morin, E., Mugnai, A., and Llasat, M. C.: Predicting the potential for lightning activity in Mediterranean storms based on the Weather Research and Forecasting (WRF) model dynamic and microphysical fields, J. Geophys. Res.-Atmos., 115, D04205, doi:10.1029/2008JD010868, 2010.

Ziegler, C. L., MacGorman, D. R., Dye, J. E., and Ray, P. S.: A model evaluation of noninductive graupel-ice charging in the early electrification of a mountain thunderstorm, J. Geophys. Res., 96, 12833–12855, 1991.

7

BINARY: an optical freezing array for assessing temperature and time dependence of heterogeneous ice nucleation

C. Budke and T. Koop

Faculty of Chemistry, Bielefeld University, Universitätsstraße 25, 33615 Bielefeld, Germany

Correspondence to: T. Koop (thomas.koop@uni-bielefeld.de)

Abstract. A new optical freezing array for the study of heterogeneous ice nucleation in microliter-sized droplets is introduced, tested and applied to the study of immersion freezing in aqueous Snomax® suspensions. In the Bielefeld Ice Nucleation ARraY (BINARY) ice nucleation can be studied simultaneously in 36 droplets at temperatures down to $-40\,^\circ$C (233 K) and at cooling rates between 0.1 and $10\,\mathrm{K\,min^{-1}}$. The droplets are separated from each other in individual compartments, thus preventing a Wegener–Bergeron–Findeisen type water vapor transfer between droplets as well as avoiding the seeding of neighboring droplets by formation and surface growth of frost halos. Analysis of freezing and melting occurs via an automated real-time image analysis of the optical brightness of each individual droplet. As an application ice nucleation in water droplets containing Snomax® at concentrations from $1\,\mathrm{ng\,mL^{-1}}$ to $1\,\mathrm{mg\,mL^{-1}}$ was investigated. Using different cooling rates, a small time dependence of ice nucleation induced by two different classes of ice nucleators (INs) contained in Snomax® was detected and the corresponding heterogeneous ice nucleation rate coefficient was quantified. The observed time dependence is smaller than those of other types of INs reported in the literature, suggesting that the BINARY setup is suitable for quantifying time dependence for most other INs of atmospheric interest, making it a useful tool for future investigations.

1 Introduction

Atmospheric ice nucleation is one of the key steps in high-altitude cloud formation and also for triggering precipitation in mixed-phase clouds (Pruppacher and Klett, 1997; Cantrell and Heymsfield, 2005; Lamb and Verlinde, 2011). Ice particles can be formed via homogeneous ice nucleation in liquid aerosol and water droplets (Koop et al., 2000; Murray et al., 2010), or via heterogeneous ice nucleation triggered by pre-existing ice nucleators (Pruppacher and Klett, 1997; Cantrell and Heymsfield, 2005; DeMott et al., 2010; Murray et al., 2012; Hoose and Möhler, 2012; Cziczo and Froyd, 2014). Both homogeneous as well as heterogeneous ice nucleation processes do occur, and various approaches of parameterizing them in atmospheric models have been described. One strategy for improving the description of ice nucleation in cloud models is the elucidation of the responsible mechanisms and their physical dependencies in laboratory experiments. For example, such laboratory data can then serve as a basis for physically consistent parameterizations for heterogeneous ice nucleation that can be incorporated into process models (Hoose and Möhler, 2012). In mixed-phase clouds heterogeneous immersion mode freezing is thought to be a relevant process (Pruppacher and Klett, 1997; Lohmann and Diehl, 2006). But time dependence of immersion freezing is often poorly represented or not included at all in cloud models, in order to reduce model complexity (Ervens and Feingold, 2012). It is well accepted that homogeneous ice nucleation is a time-dependent stochastic process which can be described by the formation of an ice embryo with critical size, whose probability of forming increases with time (Pruppacher and Klett, 1997; Murray et al., 2010; Riechers et al., 2013). There is, however, an ongoing debate on whether heterogeneous ice nucleation in the immersion mode is only temperature dependent (i.e., a singular process) or both temperature and time dependent (i.e., a stochastic process) (Vali, 2014). In a stochastic process, the probability of the occurrence of a nucleation event increases exponentially with time at a rate that depends on temperature (e.g., Bigg, 1953; Vonnegut and Baldwin, 1984; Pruppacher and Klett, 1997; Vali, 1994, 2014). In contrast, in a singular process no such time dependence of ice nucleation exists, because the probability of ice nucleation switches instantaneously from 0 to 1 at a deterministic temperature that depends on the IN (e.g., Vali and Stansbury, 1966). Moreover, another so-called modified singular process has been proposed that describes the overall process as temperature dependent with a small stochastic variation around the deterministic temperature (Vali, 2008).

Several studies have employed experimental data and model calculations to show very little time dependence, thus justifying the use of the time-independent singular approach (Kulkarni et al., 2012; Welti et al., 2012; Wright and Petters, 2013; Ervens and Feingold, 2013; Vali, 2014). But it has been also suggested that any existing small time dependence should not be neglected if a more accurate description of heterogeneous nucleation is to be achieved in models (Barahona, 2012; Knopf and Alpert, 2013; Vali and Snider, 2014). One drawback from which many experimental techniques suffer is the fact that cooling rates or the rates by which the supersaturation changes can be varied over a small range only, thus limiting their sensitivity to distinguish between the different approaches (see, e.g., the discussion in Niedermeier et al., 2010; Lüönd et al., 2010).

Several instrumental techniques are available for the determination of temperature and time dependence of heterogeneous ice nucleation (see, e.g., Murray et al., 2012; Hoose and Möhler, 2012). One of the commonly used methods is based on an early description of a drop-freezer apparatus developed by Vali and Stansbury (1966) and Vali (1971a). In the original setup, microliter-sized droplets are pipetted onto a substrate that is placed on a thermoelectric cooler. The droplets are separated by about 1 cm from each other, but no other precautions are taken to avoid a Wegener–Bergeron–Findeisen type process, i.e., the growth of frozen droplets at the expense of remaining supercooled liquid ones by water vapor transfer. In addition, the latent heat release during freezing may cause the formation of frost halos that subsequently may grow and expand around the frozen droplets by Wegener–Bergeron–Findeisen type water vapor transfer (Jung et al., 2012; Welz, 2013). This frost halo growth bears the risk of seeding neighboring droplets and thus biasing the recorded ice nucleation temperatures (Jung et al., 2012; Welz, 2013). To circumvent these problems and to minimize evaporation, droplets are often covered with an oil film (Bigg, 1953; Hoffer, 1961; Murray et al., 2011; Pummer et al., 2012; Wright and Petters, 2013). But the use of such emulsion-type samples may lead to alternative problems because some ice nucleators (INs) such as pollen or fungal spores may have an affinity to the hydrophobic phase. When a part of the dispersed IN material is lost to the oil phase, an overestimation of IN concentration may result, leading to an underestimation of IN activity. Another uncertainty arises if surfactants are used for the stabilization of emulsions, because surfactants may also influence the investigated INs at the water–oil interface (Pummer et al., 2012), particularly when the INs reside predominantly at the droplet surface.

A very recent instrument development of Stopelli et al. (2014) uses sealable tubes to prevent evaporation and cross-contamination of the investigated suspensions. The device bears an additional advantage over previous tube-based ice nucleation devices (e.g., Barlow and Haymet, 1995; Heneghan et al., 2002) in that it allows simultaneous investigation of multiple samples rather than focusing on numerous freeze–thaw repeats of a limited number of samples.

Another improvement is the nature of the substrate on which the droplets are positioned. In the original drop freezer, water droplets were pipetted onto an oil-covered sheet of aluminum foil (Vali and Stansbury, 1966). More recent techniques make use of advanced chemical approaches to hydrophobize glass or quartz substrates by a self-assembled monolayer (e.g., using a halogenated silane), which does not affect the mechanical or thermal properties of the substrate (Koop et al., 1998; Salcedo et al., 2000; Knopf and Lopez, 2009; Murray et al., 2010; Iannone et al., 2011).

Some of the devices introduced in the past for the analysis of heterogeneous ice nucleation employ rather large sample volumes (tens to hundreds of microliters) in order to be able to further analyze the ice-nucleating substances, for long-term IN sample storage and an associated investigation of ageing effects, or to ease the detection of freezing (Barlow and Haymet, 1995; Attard et al., 2012; Stopelli et al., 2014). Others have focused on rather small (often emulsified) samples (picoliters to nanoliters) in order to minimize or exclude the unwanted effects of ice-nucleating impurities contained in the water for preparing the suspensions under scrutiny (Koop and Zobrist, 2009; Murray et al., 2010, 2011; Pummer et al., 2012; Wright and Petters, 2013; Atkinson et al., 2013). Herein we use intermediate volume samples of $1\,\mu L$ which are easy to prepare and are not subject to concentration uncertainties owing to water evaporation during sample preparation and storage. For such volumes heterogeneous ice nucleation from impurities contained in the "pure" water usually occurs at temperatures below $-20\,°C$ (253 K). Therefore, such samples are applicable for ice nucleation studies over the important temperature range of mixed-phase clouds between $-20\,°C$ (253 K) and $0\,°C$ (273 K). At lower temperatures, corrections for the effects of IN impurities contained in "pure" water must be employed.

In the following, we introduce a new droplet freezing assay, in which the separation of the investigated water droplets is accomplished by a polymer spacer. This spacer encloses each of a total of 36 microliter-sized droplets individually without direct contact to any of them. Moreover, we have developed an automated system for analyzing the freezing temperature of each drop. Finally, the device allows for an accurate determination of ice nucleation temperatures over a large range of cooling rates from 0.1 to 10 K min^{-1}. We chose Snomax®, a commercially available ice inducer used in snow cannons, as a test substance for the investigation of heterogeneous ice nucleation and an assessment of its temperature and time dependence.

2 Theoretical background for freezing analysis

2.1 Singular (deterministic) data analysis

The temperature dependence of ice nucleation induced by singular ice nucleators can be described in terms of $n_{\mathrm{m}}(T)$, the cumulative number of ice nucleators per mass of Snomax® as a function of temperature. $n_{\mathrm{m}}(T)$ can be obtained from droplet freezing array data by analyzing the frozen fraction f_{ice} from the cumulative number of frozen droplets $n_{\mathrm{ice}}(T)$ and the total number of droplets n_{tot} of a particular Snomax® concentration (Vali, 1971b):

$$f_{\mathrm{ice}}(T) = \frac{n_{\mathrm{ice}}(T)}{n_{\mathrm{tot}}} = 1 - e^{-K(T)V}. \tag{1}$$

From Eq. (1) the cumulative ice nucleator concentration $K(T)$ can be deduced, which is typically referred to as the active site number density per unit droplet volume V. Because in the experiments described below we know the mass concentration C_{m} of Snomax® and the volume V of the droplets, $K(T)$ can be converted into the active site density per unit mass $n_{\mathrm{m}}(T)$:

$$n_{\mathrm{m}}(T) = \frac{K(T)}{C_{\mathrm{m}}} = n_{\mathrm{s}}(T) \cdot S' = n_{\mathrm{n}}(T) \cdot N'. \tag{2}$$

Likewise the active site density per unit surface area $n_{\mathrm{s}}(T)$ or per particle number $n_{\mathrm{n}}(T)$ can be deduced if the specific surface area S', i.e., the surface area per sample mass, or the specific particle number N', i.e., the number of particles per sample mass, respectively, are known from independent analysis (Fletcher, 1969; Connolly et al., 2009; Murray et al., 2012; Niemand et al., 2012).

The value of the specific surface area S' of Snomax® can be derived as follows. For the specific particle number of cells in Snomax® we used the value of $N' = 1.4 \times 10^9$ mg^{-1} determined by Wex et al. (2015) using multiple instruments including the BINARY setup presented here. Furthermore, Wex et al. (2015) investigated the size of the particles/cells suspended in freshly prepared Snomax® suspensions by dynamic light scattering, resulting in a mean equivalent hydrodynamic radius of $r_{\mathrm{h}} = 0.5\,\mu$m. Accordingly, we assumed that each cell has a surface area equivalent to that of a sphere with a radius of $0.5\,\mu$m, i.e., a surface area of $3.14\,\mu\mathrm{m}^2$ per cell. These numbers result in a value of $S' = 44$ cm^2 mg^{-1}, which was used here.

2.2 Time-dependent (stochastic) data analysis

A time dependence of ice nucleation induced by ice nucleators can be investigated by performing freezing experiments at different cooling rates in order to determine the nucleation rate $R(T)$. For such experiments, temperature is divided into equally spaced intervals $\Delta T = T_1 - T_2 = 0.1$ K and all data are binned into these temperature intervals with interval temperature $T = T_1 - 0.5 \cdot \Delta T$. When droplets are cooled at a constant cooling rate β, the nucleation rate $R(T)$ at the interval temperature T can be derived from the number of droplets nucleated in that interval $\Delta n_{\mathrm{ice}}(T)$ (Vali, 1994; Koop et al., 1997; Zobrist et al., 2007):

$$R(T) = \frac{\Delta n_{\mathrm{ice}}(T)}{t_{\mathrm{tot}}(T)} \tag{3}$$

$$= \frac{\Delta n_{\mathrm{ice}}(T) \cdot \beta}{\Delta T \cdot \left(n_{\mathrm{liq}}(T_1) - \Delta n_{\mathrm{ice}}(T)\right) + \sum\limits_{j=1}^{\Delta n_{\mathrm{ice}}(T)} (T_1 - T_{\mathrm{nuc},j})}, \tag{4}$$

where $j = 1, \ldots, \Delta n_{\mathrm{ice}}(T)$ are the nucleation events in individual droplets occurring at temperature $T_{\mathrm{nuc},j}$ within the interval ΔT. With the simplification $T_{\mathrm{nuc},j} \approx T_1 - 0.5 \cdot \Delta T$, which is a good approximation for small temperature intervals, the sum in the denominator in Eq. (4) can be approximated by $0.5 \cdot \Delta T \cdot \Delta n_{\mathrm{ice}}(T)$ and, hence,

$$R(T) \approx \frac{\Delta n_{\mathrm{ice}}(T) \cdot \beta}{\Delta T \cdot \left(n_{\mathrm{liq}}(T_1) - 0.5 \cdot \Delta n_{\mathrm{ice}}(T)\right)}. \tag{5}$$

Data of $R(T)$ can be normalized to the total surface area of ice nucleator per droplet A, which for the case of a single-component (sc) ice nucleator is equivalent to the nucleation rate coefficient $j_{\mathrm{het}}(T, \alpha)$,

$$\frac{R(T)}{A} = \frac{R(T)}{S' C_{\mathrm{m}} V} \overset{\mathrm{sc}}{=} j_{\mathrm{het}}(T, \alpha), \tag{6}$$

with the parameters S', C_{m} and V as defined above. Using the framework of classical nucleation theory (CNT) $j_{\mathrm{het}}(T, \alpha)$ can be described as (Pruppacher and Klett, 1997)

$$j_{\mathrm{het}}(T, \alpha) = \frac{kT}{h} \exp\left(-\frac{\Delta F_{\mathrm{diff}}(T)}{kT}\right)$$
$$\cdot n \exp\left(-\frac{\Delta G(T)\varphi(\alpha)}{kT}\right). \tag{7}$$

Here k is the Boltzmann constant, h is the Planck constant, T is absolute temperature and n is the number density of water molecules at the IN–water interface. $\Delta F_{\mathrm{diff}}(T)$ and $\Delta G(T)$ are the diffusion activation energy of a water molecule crossing the water–ice embryo interface and the Gibbs free energy for critical ice embryo formation without the presence of a heterogeneous IN, respectively. In the presence of an ice nucleus $\Delta G(T)$ is modified by the compatibility function $\varphi(\alpha)$:

$$\varphi(\alpha) = \frac{1}{4}(2 + \cos \alpha)(1 - \cos \alpha)^2, \qquad (8)$$

where the parameter α represents a hypothetical effective contact angle between the forming ice embryo and the IN in water, which can vary between 0 and 180°. An effective contact angle of $\alpha = 0°$ implies a perfect IN leading to $\varphi = 0$ and a negligible Gibbs free energy term (equivalent to seeding by a mother crystal). An effective contact angle of $\alpha = 180°$ leads to $\varphi = 1$, and the Gibbs free energy term is not reduced, implying a maximally poor IN; thus, nucleation occurs at the rate of homogeneous ice nucleation. For our analysis of $j_{het}(T, \alpha)$ from experimental data we use α as the only free parameter when fitting Eq. (7) to the data. Temperature-dependent parameterizations for all other quantities were obtained from Zobrist et al. (2007).

2.3 Quantification of time dependence

In a very recent study Herbert et al. (2014) proposed the following equation for the description of cooling rate dependence of a single IN (applied to $T_{f,50}$ values here):

$$T_{f,50}(\beta_2) - T_{f,50}(\beta_1) = \Delta T_{f,50} = \frac{1}{\lambda} \cdot \ln\left(\frac{\beta_1}{\beta_2}\right). \qquad (9)$$

$T_{f,50}(\beta_2)$ and $T_{f,50}(\beta_1)$ are the median freezing temperatures at cooling rates β_2 and β_1, respectively, and λ is a constant. Herbert et al. (2014) point out that, for a single-component IN following stochastic (i.e., non-singular) behavior, λ is the temperature dependence of the heterogeneous ice nucleation rate coefficient of the IN: $\lambda = -d\ln(j_{het})/dT$. We note that an equation similar to Eq. (9) was originally introduced by Vali and Stansbury (1966) for the mean freezing temperature, i.e., $\overline{T}_f(\beta_2) - \overline{T}_f(\beta_1) = \xi \cdot \ln\left(\frac{\beta_1}{\beta_2}\right)$. Ignoring any difference between $\overline{T}_f$ and $T_{f,50}$ (which occurs, e.g., for a T_f distribution that is constant with changing cooling rate) implies that the constant ξ is related to λ via $\xi = \lambda^{-1}$. For the singular case $\xi_{si} = 0$ and $\lambda_{si} = \infty$ by definition, and a small ξ/large λ value indicates a behavior that is close to singular; i.e., nucleation shows a less-pronounced stochastic time dependence. Moreover, one can define the temperature dependence of the normalized freezing rate $\omega = -d\ln(R/A)/dT$ and note that $\omega = \lambda$ for a single-component IN and $\omega < \lambda$ for a multi-component IN (Vali, 2014; Herbert et al., 2014).

3 Experimental setup

The optical freezing apparatus introduced here, which we term BINARY (Bielefeld Ice Nucleation ARraY), consists of a 6×6 array of individual microliter-sized droplets positioned on a thin hydrophobic glass slide (Fig. 1a). The droplets are separated from each other by a soft poly-dimethylsiloxane (PDMS) spacer, and the resulting compartments are sealed at the top with another glass slide (Fig. 1b). The PDMS spacer is fabricated from a $10:1$ mixture of base polymer to curing agent (Sylgard® 184, Dow Corning) poured onto a mold made from aluminum that was custom-designed to represent the compartment array.

The droplet separation into individual compartments prevents a Wegener–Bergeron–Findeisen process, in which frozen droplets grow at the expense of unfrozen supercooled liquid droplets due to the vapor pressure difference between ice and supercooled liquid water (Murphy and Koop, 2005). This process poses a particular problem to droplet arrays operated at small cooling rates or employing stepwise cooling and has been noted frequently to influence and corrupt ice nucleation measurements (Welz, 2013; Stopelli et al., 2014; O'Sullivan et al., 2014). In addition, the probability of heterogeneous ice nucleation at the lower glass surface is minimized by the hydrophobicity of the glass (see discussion above). Water droplets positioned on the silanized glass show contact angles of about 90–100° (Remmers, 2012). Hence, the shape of the droplets investigated below is well approximated by a hemisphere.

Snomax® suspensions were prepared by mixing a predetermined mass of dry material with the appropriate volume of freshly double-distilled water. Individual droplets of $1\,\mu L$ volume were sampled from the suspension with a micropipette (volume inaccuracy $\leq 3\,\%$) and positioned individually on the glass surface in each compartment. We note that the use of smaller (e.g., $\sim 0.5\,\mu L$) and larger (e.g., $\sim 5\,\mu L$) droplets is also possible, but only $1\,\mu L$ droplets were used in the freezing experiments described below.

The sample array is placed onto a Peltier cooling stage (Fig. 1c), which is part of a commercially available cooling stage system (Linkam LTS120). Good thermal contact between the Peltier stage and the lower glass surface of the sample array is achieved by pressing the upper glass slide towards the stage with four fixing screws, one at each corner. When the Peltier stage is connected to a heat sink bath at $5\,°C$ (278 K), the sample array can be cooled to $-40\,°C$ (233 K) at cooling rates between 0.1 and $10\,K\,min^{-1}$. The cover of the cooling stage was modified to consist of a larger opening sealed with a thin glass window such that the resulting larger field of view of $40\,mm \times 40\,mm$ allowed for the simultaneous observation of all droplets of the sample array. In addition, small arrays of cold-light white LEDs were fixed to the top of the cooling chamber to yield the proper contrast for identifying the phase state of the droplets; see below. Finally, purging of dry N_2 gas into the chamber as well as onto the top window prevents dew and frost formation from laboratory humidity during cooling.

A LabVIEW™ virtual instrument is used to control the temperature of the Peltier stage and to analyze in real time the digital images obtained by a CCD camera (QImaging MicroPublisher 5.0 RTV). The images are recorded and an-

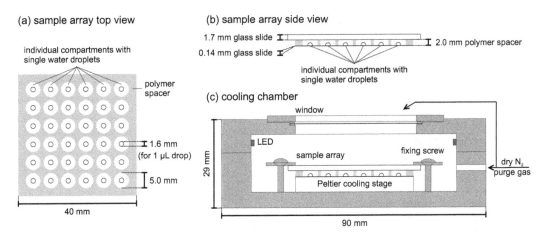

Figure 1. Schematic of the Bielefeld Ice Nucleation ARraY (BINARY) setup. **(a)** Top view of the 6×6 droplet array. The droplets are separated from each other by a polymer spacer creating individual compartments. **(b)** Side view showing the sealing of the compartments by top and bottom glass slides. **(c)** Position of the sample array on the Peltier cooling stage inside the cooling chamber.

alyzed at a frequency that depends upon the experimental cooling rate: three successive images are analyzed per 0.1 K temperature interval, i.e., one image every 0.03 K. For example, the corresponding time interval $|\Delta t_{j-1,j}|$ between successive images $j-1$ and j at a cooling rate of 1 and 5 K min^{-1} is 2 and 0.4 s, respectively. Moreover, every third image is stored in digital format, allowing for later re-examination or re-analysis. These images have a temperature resolution of 0.1 K.

Ice nucleation is determined optically based on the change in droplet brightness when the initially transparent liquid droplets become opaque upon freezing (see Fig. 2a and the video of this experiment available in the Supplement). Since we observe the droplets in reflection mode (not in transmission), in the images liquid droplets appear dark, and frozen droplets appear bright as they scatter light into the observation light path. This change in brightness is maximized by illuminating the droplets by LEDs at a low sideway angle from the top (see Fig. 1c) and also by the reflective top surface of the Peltier stage. The 8 bit mean gray value, gv, (ranging from gv = 0 for black to gv = 255 for white) is determined for each compartment/droplet i in every image j. The difference in gv between successive images and, hence, temperatures $\Delta gv_{i,j}(T) = gv_{i,j} - gv_{i,j-1}$ is then used to determine droplet freezing and melting.

Figure 2 shows an example experiment of water droplets containing Snomax$^{\circledR}$ at a concentration of 0.1 μg drop^{-1}. In panel a three images are shown of the entire array during cooling. At $-3.5\,°C$ (269.6 K) all droplets are still liquid as recognized by their dark appearance, and at $-5.5\,°C$ (267.6 K) all droplets are frozen and bright. At the intermediate temperature of $-4.5\,°C$ (268.6 K) some bright spots indicate frozen or currently freezing droplets. In panel b a more detailed analysis is exemplified for the droplet shown in the yellow box in panel a. The droplet is first cooled and subsequently reheated at rates of 1 K min^{-1}. The mean gray

value gv for the droplet is determined from all pixels contained in the yellow box. In the beginning the gray value increases only slightly during cooling, due to condensation of water vapor contained in the compartment onto the bottom glass plate.[1] Ice nucleation in the droplet occurs at $-3.9\,°C$ (269.3 K), resulting in a steep increase of the gray value from about 30 to nearly 70 at $-4.2\,°C$ (269.0 K) in the few subsequent images. While the largest gray value change Δgv between two consecutive images is about 10 at a temperature slightly below $-4\,°C$ (269 K; see Fig. 2c for comparison), nucleation and freezing is detected already a few images before. We set a threshold value of $\Delta gv > 1$ for the automatic attribution of ice nucleation, since all other Δgv values during the entire cooling procedure are significantly smaller than 1 at a typical noise level of $\Delta gv \lesssim \pm 0.2$ (Fig. 2c).

Following droplet freezing the previously condensed liquid water then evaporates and freezes onto the frozen droplet, leading to the formation of a frost halo and a slow decrease in gv to about 62 in this particular case (Fig. 2b). After a temperature of $-10\,°C$ is reached, the sample is reheated. The gray value slightly decreases further since the polycrystalline ice slowly recrystallizes, thus reducing the amount of

---

[1] The condensing water originates from the humidity inside each compartment from the vapor pressure of the droplet itself. At the start of the experiment the relative humidity in each compartment is 100 %, and, upon cooling, some of this humidity condenses onto the cold glass slide. These droplets are much smaller than the investigated droplets and, thus, normally freeze at much colder temperature, i.e., at about the homogeneous ice nucleation limit of supercooled water. Therefore, the condensed droplets do not affect the heterogeneous ice nucleation processes in the microliter droplets studied here. Only if heterogeneous ice nucleation is triggered in the condensed droplets (e.g., by a surface impurity/irregularity) might they subsequently seed the larger droplets. This was observed to occur only rarely at the lowest cooling rates, and, accordingly, these data points were excluded from the analysis.

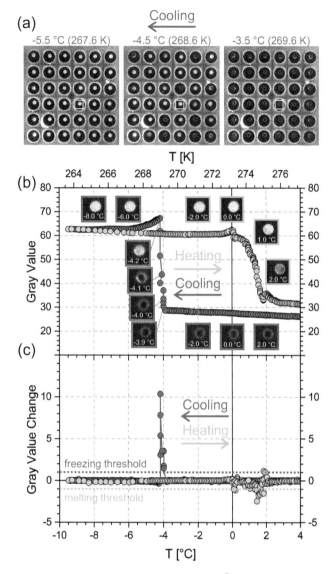

Figure 2. Example experiment with Snomax®-containing droplets $(0.1\,\mu g\,drop^{-1})$ describing the automatic detection of nucleation events by the change in brightness during freezing. A video of this experiment is available in the Supplement. **(a)** Image series of the 6×6 droplet array during cooling. **(b)** Measured gray value of the compartment/droplet indicated by the yellow box in panel **(a)** during cooling (red) and heating (green). Freezing and melting start at $-3.9\,°C$ (269.3 K) and $0.0\,°C$ (273.2 K), respectively. **(c)** Plot of the change in gray value between successive images showing peaks at the phase transition points. Threshold values of ±1 for the automatic attribution of freezing and melting are indicated by dashed lines.

ice facets capable of scattering the LED light. Finally, gv increases just before reaching the melting point due to formation of liquid water films in the grain boundaries of the frozen droplet; thus LED light is reflected slightly more efficiently. The onset of melting at $0\,°C$ (273 K), most easily observed at the halo, finally results in a decrease of gv and hence a lo-

cal minimum in Δgv; see Fig. 2c. We set a threshold value of $\Delta gv < -1$ for automatic attribution of melting. We note that at a heating rate of $1\,K\,min^{-1}$ the melting of the entire droplet is complete at about $2\,°C$ (275 K).

The analysis described in Fig. 2 is performed automatically for all droplets of a particular array. The nucleation temperatures thus obtained for each droplet undergo a correction according to a temperature calibration that is outlined in the next section.

4   Temperature calibration

The experiment discussed in the previous section was performed at a constant cooling rate of $1\,K\,min^{-1}$. Investigation of time dependence of heterogeneous ice nucleation in such constant cooling rate experiments requires performing several experiments at different cooling rates (see, e.g., Herbert et al., 2014). Therefore, we conducted a comprehensive calibration exercise that accounts for variable cooling rates in BINARY. The calibration was performed using five reference phase transitions in the temperature range of interest from -37 to $0\,°C$ (236 to 273 K; see Table A1 and Fig. A1 in the Appendix) and for heating rates from 0.1 to $10\,K\,min^{-1}$. The rate calibration was conducted in the heating mode because superheating of a crystal above its melting point is usually negligible. In contrast, supercooling of a liquid below the melting point often occurs readily, and even minor supercooling would bias a calibration in cooling mode (Sarge et al., 2000; Della Gatta et al., 2006). For the solid–liquid phase transition of tridecane the rate dependence of cooling vs. heating was investigated. Although crystal nucleation temperatures of tridecane droplets in the size range from 0.1 to $1.5\,\mu L$ scattered significantly more – by about 0.8 K between the 25th to 75th percentiles – than those of the melting temperatures during calibration – about 0.2 K – the observed dependence of nucleation temperatures with increasing cooling rate was similar to that of the melting temperatures with increasing heating rate, which were ~ 0.02 and $\sim 0.05\,min$, respectively.

For the calibration procedure the reference substances were sprayed onto the hydrophobic glass slide, resulting in droplets $\sim 0.6\,mm$ in size (estimated droplet volume about $0.05\,\mu L$). The onset of melting and the subsequent, almost-immediate full melting in such small droplets results in a steeper Δgv, which is easier to detect and, hence, more accurate than in larger droplets, which often melt entirely only with measurable delay; see Fig. 2. The threshold value for automatic detection of the phase transitions was adjusted for each calibration substance, in particular for the solid–solid phase transitions which involve smaller Δgv values. The phase transition temperatures obtained for the reference substances were analyzed by separating the absolute temperature offset of the sample array at a heating rate of 0 (i.e., at static conditions) from its heating rate dependence due to

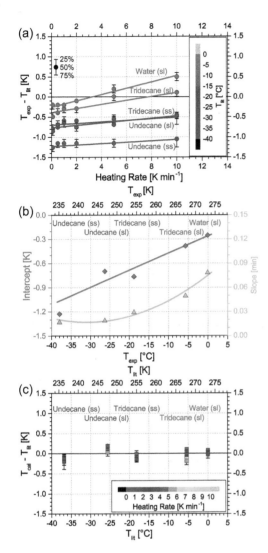

Figure 3. Temperature calibration. (a) Difference between the experimentally determined phase transition temperatures of water, tridecane and undecane droplets and the corresponding literature values as a function of heating rate. Linear fits to the data are indicated by the solid lines. (b) Intercepts (left, red) and slopes (right, green) of the linear fits shown in (a) above. The experimentally determined data (symbols) are fitted by a linear function (intercepts, red line) and by a second-order polynomial (slopes, green line). (c) Residual difference between the calibrated phase transition temperatures and the literature values after calibration.

thermal lag, as recommended for the calibration of thermal devices (Sarge et al., 2000; Della Gatta et al., 2006; Riechers et al., 2013). Figure 3a shows the difference between the experimentally observed temperatures T_{exp} and the literature values T_{lit} as a function of heating rate. Points represent the median values, and error bars indicate the 25th and 75th percentiles of the data from different compartments and multiple cooling–heating cycles. Each individual reference substance's data set was fitted linearly. The absolute temperature deviation at a heating rate of 0 corresponds to the intercept of

such a fit and the heating rate dependence to the slope. The fitting parameters obtained are shown in Fig. 3b as a function of T_{exp} (intercept: red diamonds and left axis; slope: green triangles and right axis). These intercepts are well represented by a linear fit (red line) and the slopes by a second-order polynomial (green line). These fits were then applied to the raw data. The remaining temperature uncertainty after this calibration step, which is the absolute difference between the calibrated temperatures T_{cal} and the literature values T_{lit}, is shown in Fig. 3c. The squares represent the median values and the error bars the 25th and 75th percentiles for each calibration substance. We note that $|T_{cal} - T_{lit}|$ is smaller than 0.3 K for 97 % of all individual data points, indicating the quality of the calibration procedure.

5   Results

We put the new BINARY setup to the test using Snomax® as a well-studied ice-nucleating substance (Maki et al., 1974; Vali et al., 1976; Ward and DeMott, 1989; Turner et al., 1990; Wood et al., 2002; Möhler et al., 2008; Hartmann et al., 2013; Stopelli et al., 2014). Snomax® is a commercial product containing freeze-dried nonviable bacterial cells from *Pseudomonas syringae*, which are known to be active INs at high temperature. First, experiments with droplets of 1 µL volume were investigated at a cooling rate of 1 K min^{-1}. Snomax® concentrations were varied over 6 orders of magnitude between 1 ng mL^{-1} and 1 mg mL^{-1}, corresponding to a total mass between 1 pg per droplet and 1 µg per droplet, respectively. Usually, 108 droplet freezing events (from 3 × 36 droplets) were analyzed at each concentration.

The results of these experiments are shown in Fig. 4 by plotting $n_m(T)$, the cumulative number of ice nucleators per µg of Snomax®, as a function of temperature. Two particularly strong increases in $n_m(T)$ are observed, one at about −3.5 °C (269.6 K) ±0.5 K and one at −8.5 °C (264.6 K) ±0.5 K, indicating the presence of two distinct classes of ice nucleators with different activation temperatures. The two plateaus at temperatures just below each increase of $n_m(T)$ in Fig. 4 arise when no INs active at these temperatures are present in the investigated suspensions. The $n_m(T)$ values of the plateaus differ by about 3 orders of magnitude, from which we infer that the two classes of Snomax® INs occur at a number ratio of about 1 to 1000 in our samples. The active site densities per cell $n_n(T)$ shown in Fig. 4 on the right axis were calculated using the specific particle number of cells in Snomax® determined by Wex et al. (2015); see Sect. 2.1.

It is already known that different types of INs or aggregates of INs are responsible for ice nucleation induced by *Pseudomonas* and other ice-nucleating bacteria in different temperature ranges (Yankofsky et al., 1981; Turner et al., 1990; Hartmann et al., 2013). For example, Turner et al. (1990) identified three classes of INs on the basis of the temperature at which they triggered ice nucle-

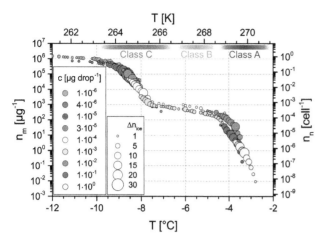

Figure 4. Experimentally determined active site density per unit mass of Snomax® $n_m(T)$ vs. temperature. Symbol colors indicate data from droplets with different Snomax® concentrations; symbol size indicates the number of nucleating droplets per temperature interval. The temperature range for different classes of INs according to the definition by Turner et al. (1990) are also indicated by the colored bars.

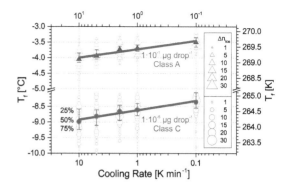

Figure 5. Freezing temperatures of water droplets containing Snomax® at concentrations of $1 \times 10^{-5}\,\mu g\,drop^{-1}$ (red) and $1 \times 10^{-1}\,\mu g\,drop^{-1}$ (blue) as a function of experimental cooling rate. The number of nucleation events at individual temperatures is indicated by the size of the open symbols. Solid symbols indicate the median freezing temperature ($T_{f,50}$), with error bars representing the 25th and 75th percentiles. Solid lines are linear fits to the solid symbols; see text.

ation at various conditions: class A at high temperature ($\gtrsim -4.5\,°C \simeq 269\,K$), class B in the intermediate range (approx. -4.5 to $-6.5\,°C \simeq 269$ to $267\,K$) and class C at lower temperature ($\lesssim -6.5\,°C \simeq 267\,K$); see the colored bars at the top of Fig. 4. According to this definition the data shown in Fig. 4 suggest that our sample contained class A and class C ice-nucleating proteins or protein complexes. In contrast, Turner et al. (1990) identified Snomax® to contain INs of class A and class B, but none of class C. This difference to our study may result from varied growth or storage conditions of different Snomax® samples and is in line with their proposal that class C INs can develop into class B (and subsequently into class A) INs by accumulation of larger ice-nucleating protein complexes in the cell membranes. In fact, our sample may also contain class B INs, but the low number of freezing events in the corresponding temperature range of -4.5 to $-6.5\,°C$ (269 to 267 K) indicates that there are fewer INs of class B than there are of class A, the latter of which trigger ice nucleation already at higher temperature. As mentioned above, in our sample the more active but less abundant class A ice nucleators induce freezing at about $-3.5\,°C$ (269.6 K) $\pm 0.5\,K$, and they dominate the freezing of droplets at a Snomax® concentration of $1 \times 10^{-1}\,\mu g\,drop^{-1}$. The less active but more abundant INs of class C nucleate ice at about $-8.5\,°C$ (264.6 K) $\pm 0.5\,K$. At a Snomax® concentration of $1 \times 10^{-5}\,\mu g\,drop^{-1}$ almost all droplets contain such INs of class C, but practically none of class A anymore.

The analysis shown in Fig. 4 is based on the singular approach which assumes that the results are independent of time and, hence, cooling rate. With the new BINARY setup we were able to challenge this assumption by performing experiments at different cooling rates in the range from 0.1 to

$10\,K\,min^{-1}$. In particular the cooling rate dependence of the two IN classes discussed above was investigated. Figure 5 shows a decrease in the median ice nucleation temperature $T_{f,50}$ with increasing cooling rate for both classes of INs. The $T_{f,50}$ values are represented by filled symbols with error bars indicating the 25th and 75th percentiles. At the indicated concentrations the difference between the $T_{f,50}$ values at 10 and at $0.1\,K\,min^{-1}$ is about 0.6 K for both classes of INs (0.55 K for class A and 0.64 K for class C). These values are small but larger than our temperature uncertainty ($\pm 0.3\,K$ on a 2σ level). While this degree of time dependence is probably not of atmospheric importance, we analyze and quantify it in more detail below for two reasons: first, it is interesting from a physical chemistry viewpoint regarding the fundamental process of heterogeneous ice nucleation; second, such an analysis may help in characterizing the ability and limitations of the BINARY device for measurements of time dependence of INs more generally.

The above analysis suggests a very small time dependence of Snomax®-induced ice nucleation. Nevertheless, we may expect to see deviations in the active site density obtained at different cooling rates. Figure 6a shows $n_m(T)$ at the two investigated Snomax® concentrations at different cooling rates. Clearly, there is a systematic trend towards lower $n_m(T)$ with larger cooling rate; i.e., the individual $n_m(T)$ curves are shifted to lower temperature as the cooling rate increases. This analysis supports the interpretation of a time dependence of ice nucleation induced by the two class A and class C Snomax® INs. Therefore, we analyze the data shown in Fig. 6a with the stochastic approach by determining the nucleation rate $R(T)$; see Eq. (6). Figure 6b shows $R(T)$ for both classes of INs. Now the data points obtained from the different cooling rates converge onto a single line, with data obtained at larger cooling rate (yellow) representing larger values of R than those obtained at a lower cooling rate (blue). These data of $R(T)$ can be normalized to the total surface area per droplet A, using the specific surface area $S' = 44\,\mathrm{cm^2\,mg^{-1}}$ of Snomax® from Wex et al. (2015); see Sect. 2.1. For a single-component ice nucleator, $R(T)/A$ is equivalent to nucleation rate coefficient $j_{\mathrm{het}}(T)$; see Eq. (6). The results of the analysis according to Eq. (6) are the data points plotted in Fig. 6c, which shows that $R(T)/A$ increases by about 3 orders of magnitude for a reduction in temperature of about 2 K for class C (circles) and by about 4 orders of magnitude for a reduction of about 1 K for class A (triangles).

The lines in Fig. 6c are fits to these data using the framework of classical nucleation theory (CNT); i.e., they represent $j_{\mathrm{het}}(T, \alpha)$ fits in which the effective contact angle α was either constant or allowed to have a linear temperature dependence. Figure 6c shows that the measured data are not well described by CNT when using constant effective contact angle values of $\alpha = 23.9°$ and $\alpha = 35.3°$ for class A and C, respectively (gray lines). Allowing for a linear temperature dependence of α, a much better fit of CNT to the data results (red and blue lines). The corresponding linear equations for the temperature-dependent effective contact angles are $\alpha(T) = 613.5 - 2.188 \cdot T$ and $\alpha(T) = 286.5 - 0.9495 \cdot T$ for class A and C, respectively. (Note that the input temperatures are in units of kelvin.) In order to visualize the effect of the steep increase of $j_{\mathrm{het}}(T)$ with decreasing temperature found for both IN classes, the parameterizations were also used to calculate the nucleation probability P_{i0} as a function of temperature and observation time (see Fig. A2 in the Appendix).

It is interesting to compare the degree of time dependence of Snomax® with that of INs of atmospheric relevance. For this purpose we follow the approach outlined in Herbert et al. (2014) to quantify the time dependence of Snomax® INs; see Sect. 2.3. Using Eq. (9) on the $T_{\mathrm{f},50}$ data shown in Fig. 5 at minimal ($0.1\,\mathrm{K\,min^{-1}}$) and maximal ($10\,\mathrm{K\,min^{-1}}$) cooling rate yields a value of $\lambda = 8.4\,\mathrm{K^{-1}}$ for the class A INs and $\lambda = 7.2\,\mathrm{K^{-1}}$ for class C INs, respectively (magenta triangles in Fig. 7; see also Table A2 in Appendix A3). Similarly, we can linearly fit the $T_{\mathrm{f},50}$ data of Fig. 5 as a function of cooling rate (orange circles in Fig. 7). We note that those two different ways for deriving λ agree well with each other; see Fig. 7 and Table A2. Moreover, we can also determine the parameter ω from fitting a straight line to $\ln(R(T)/A)$ vs. T data of Fig. 6c, resulting in the green diamonds in Fig. 7.

The nonlinear behavior of the $R(T)/A$ data in the logarithmic plot of Fig. 6c suggests that λ is not constant but instead shows a temperature dependence itself. For these reasons we finally derive $\lambda(T)$ by taking the derivative of the CNT fit resulting in the red and blue lines for $\lambda(T)$ over the investigated temperature range (Fig. 7). Clearly, the ω value is consistent with the $\lambda(T)$ value in that it represents an average value over the investigated temperature range. However, the strong difference between ω and λ obtained for the

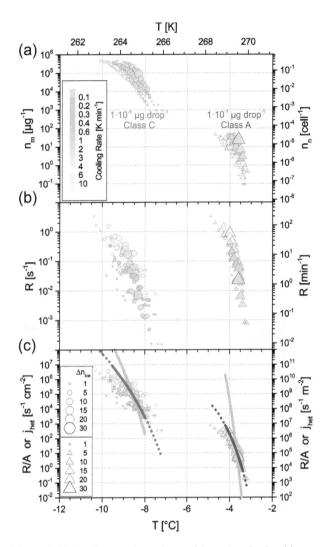

Figure 6. (a) Cooling rate dependence of the active site densities per mass ($n_{\mathrm{m}}(T)$) and per cell ($n_{\mathrm{n}}(T)$) vs. temperature for two classes of INs determined from droplets containing the indicated Snomax® concentrations. (b) Stochastic analysis of the data from (a) in terms of the nucleation rate $R(T)$. (c) Normalized nucleation rate $R(T)/A$ data derived from the $R(T)$ data in (b) following Eq. (6) (data points) and analysis of these data in terms of the heterogeneous ice nucleation rate coefficient $j_{\mathrm{het}}(T)$ using CNT with a constant effective contact angle α (gray lines) and a linear temperature dependence of α (colored lines).

class C INs and the fact that $\omega < \lambda$ indicate that class C INs may consist of a multi-component set of INs (see discussion in Herbert et al., 2014). In contrast, the good agreement between ω and λ obtained for the class A INs suggests that these are indeed single-component INs; hence, $\omega = \lambda$ in this case. We recommend a numerical value of $\omega = \lambda = 8.7\,\mathrm{K^{-1}}$ for class A Snomax® INs at the median freezing temperature of $-3.8\,°\mathrm{C}$ (269.4 K).

The λ values derived here can be compared to values for other types of INs from the literature. For example, the values for various mineral dusts (Arizona Test Dust (ATD), il-

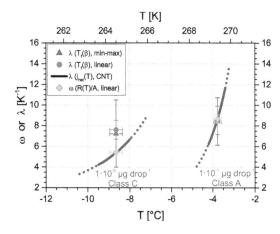

Figure 7. Value of ω and λ derived in different ways for the two indicated classes of Snomax® INs. Magenta triangle: λ obtained from Eq. (9) at the minimal (0.1 K min^{-1}) and maximal (10 K min^{-1}) cooling rate. Orange circle: λ obtained from the linear fits to the $T_{f,50}$ data shown in Fig. 5. Green diamond: $\omega = -d\ln(R/A)/dT$ obtained from a linear fit to the $R(T)/A$ data shown in Fig. 6c. Lines indicate λ derived from the derivative of the nucleation rate coefficient $-d\ln(j_{het})/dT$ based on CNT with a temperature-dependent contact angle; for details see text.

lite NX, kaolinite KGa-1b and feldspar) range between about 1 and 4.6 K^{-1} (Murray et al., 2011; Wright and Petters, 2013; Hiranuma et al., 2013; Knopf and Alpert, 2013; Vali, 2014; Herbert et al., 2014). Furthermore, the literature analysis of λ values by Herbert et al. (2014) includes also other types of INs. This comparison indicates the smallest value of $\lambda = 0.6$ K^{-1} for volcanic ash (Fornea et al., 2009; Hoyle et al., 2011; Steinke et al., 2011), and the highest value of $\lambda = 6.3$ K^{-1} for a soil sample (Vali, 2008). The small λ values for most IN compounds emphasize the stochastic nature of heterogenous ice nucleation in these cases, as small λ values correspond to a large time dependence, and vice versa. These values can be compared to homogeneous ice nucleation data in pure water at temperatures between about $-38\,°C$ (235 K) and $-35\,°C$ (238 K), which show a mean value of $\lambda = 3.4 \pm 1.2$ K^{-1} (Riechers et al., 2013).

A comparison of these literature values with those obtained here for Snomax® shows a larger λ value and implies a very small "stochasticity" of the responsible Snomax® IN moieties. To our knowledge the value of $\lambda = 8.7$ K^{-1} for the class A INs of Snomax® obtained here is the largest λ value reported to date, indicating that the time dependence of Snomax® is particularly small when compared to INs of atmospheric relevance.

6   Conclusions

Herein we introduced the novel optical freezing array BINARY for studying the temperature and time dependence of heterogeneous ice nucleation in the immersion mode. The principal advantages of the technique is prevention of a Wegener–Bergeron–Findeisen type water vapor transfer between unfrozen and frozen droplets and of a seeding of neighboring droplets by formation and surface growth of frost halos. The simultaneous study of 36 droplets and a fully automated evaluation of the ice nucleation temperature in each droplet from a real-time image analysis allows for a facile and fast accumulation of data points. Moreover, the microliter volume of the investigated droplets permits a preparation of droplets with IN concentrations varying over several orders of magnitude, thus enabling the detection of rare ice nucleators of high activity. The BINARY setup was tested by studying heterogeneous ice nucleation induced by Snomax® suspended in water droplets at various concentrations from 1 ng mL^{-1} to 1 mg mL^{-1}. Two types of INs were identified, namely class A and class C ice nucleators. Using different cooling rates, we were able to show that these INs exhibit a very small time dependence, which we quantified by analyzing the experimental data in terms of the temperature dependence λ of the ice nucleation rate coefficient. While class C Snomax® INs appear to be multicomponent, the data for class A are consistent with the behavior of a single-component IN. To the best of our knowledge, the resulting $\lambda = 8.7$ K^{-1} for the class A Snomax® INs is the largest λ value reported to date, equivalent to a particularly small time dependence. Moreover, the fact that we could quantify such a small time dependence suggests that the BINARY setup introduced here is suitable for measuring time dependence for most other INs of atmospheric interest, making it a useful tool for future analysis.

Appendix A: Additional information and data analysis

A1 Calibration literature values

For the temperature calibration described in Sect. 4 information about the literature values of the phase transitions of the reference substances were required. Most of the values are taken from Linstrom and Mallard (2014) and publications cited therein. Data which were excluded from the overall average by the National Institute of Standards and Technology (NIST) were also excluded here. In addition to the NIST data, values were taken from Parks and Huffman (1931), Messerly et al. (1967) and Mondieig et al. (2004). Figure A1 shows an overview of all data used here. A summary of the particular phase transitions and the median literature values are given in Table A1.

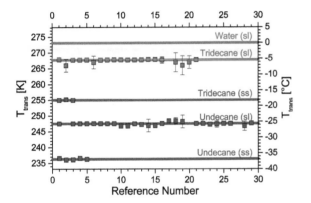

Figure A1. Literature values for the phase transitions of water, tridecane and undecane used for the temperature calibration. Symbols indicate individual literature values, and the lines represent their median used for the calibration. For data see Table A1.

Table A1. Substances and their median phase transition temperatures derived from several sources which were used in the calibration of the BINARY setup. The errors indicate the 25th and 75th percentiles.

Substance	Purity	Type of transition	Number of data points	$T_{\text{lit},50\%}$ [°C]
Water	double-distilled	solid–liquid	–	$0.00^{+0.05}_{-0.39}$
Tridecane	$\geq 99.5\%$	solid–liquid	20	$-5.41^{+0.05}_{-0.39}$
Tridecane	$\geq 99.5\%$	solid–solid	3	$-18.15^{+0.20}_{-0.00}$
Undecane	$\geq 99.8\%$	solid–liquid	26	$-25.61^{+0.01}_{-0.29}$
Undecane	$\geq 99.8\%$	solid–solid	5	$-36.85^{+0.30}_{-0.15}$

A2 Ice nucleation probability

The CNT parameterizations with a temperature-dependent $\alpha(T)$ shown as red and blue lines in Fig. 6c were used to calculate the ice nucleation probability P_{i0} as a function of time and temperature for typical experimental conditions. Here

$$P_{i0}(T, \Delta t) = \frac{\Delta n_{\text{ice}}(T, \Delta t)}{n_{\text{liq}}(T, t_1)} = 1 - e^{-R(T)\Delta t} \quad \text{(A1)}$$

represents the number of frozen droplets $\Delta n_{\text{ice}}(T, \Delta t) = n_{\text{liq}}(T, t_1) - n_{\text{liq}}(T, t_2)$ at constant temperature T in a time interval $\Delta t = t_2 - t_1$. Figure A2 shows the results for 1 microliter droplets containing Snomax® at $1 \times 10^{-1}\,\mu\text{g drop}^{-1}$ (class A) in the top panel a and for droplets containing Snomax® at $1 \times 10^{-5}\,\mu\text{g drop}^{-1}$ (class C) in the bottom panel b. The contour plots indicate very abrupt changes of the nucleation probabilities from 0 to 1 with decreasing temperature for both classes of INs. The corresponding temperature interval in which the transition occurs is smaller than about 0.5 K for time intervals of 0.1 s for class C INs. An even-smaller temperature interval of about 0.2 K results for larger time intervals and class A INs at $1 \times 10^{-1}\,\mu\text{g drop}^{-1}$. Figure A2 reinforces our interpretation provided in the Results section that the ice nucleation probability of Snomax® INs also shows a very strong temperature dependence but only a small time dependence.

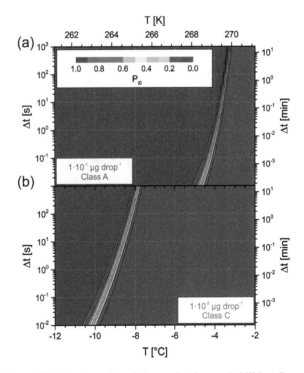

Figure A2. Contour plot of the nucleation probabilities P_{i0} as a function of time and temperature for the parametrization using CNT and temperature-dependent contact angles shown in Fig. 6c for the two investigated concentrations of $1 \times 10^{-1}\,\mu\text{g drop}^{-1}$ **(a)** and $1 \times 10^{-5}\,\mu\text{g drop}^{-1}$ **(b)**.

A3　Values for time dependence of Snomax® INs

In Table A2 we provide the results of the quantification of time dependence according to Sect. 2.3 and described in Sect. 5. The values of this analysis for two IN classes shown in Fig. 7 are provided in the following table.

Table A2. Values of λ and ω determined using different approaches for the two investigated IN classes at their median freezing temperature (given in brackets); for details see text. These data are shown also in Fig. 7.

Parameter	Method	Class C $(-8.6 \pm 0.3\,°C)$	Class A $(-3.8 \pm 0.2\,°C)$
λ	$T_f(\beta)$, min-max	$7.2 \pm 3.2\,\text{K}^{-1}$	$8.4 \pm 2.3\,\text{K}^{-1}$
λ	$T_f(\beta)$, linear	$7.6 \pm 0.9\,\text{K}^{-1}$	$8.5 \pm 1.4\,\text{K}^{-1}$
ω	$R(T)/A$, linear	$5.4 \pm 0.2\,\text{K}^{-1}$	$8.7 \pm 0.4\,\text{K}^{-1}$

Acknowledgements. The authors gratefully acknowledge funding by the German Research Foundation (DFG) through the research unit INUIT (FOR 1525) under KO 2944/2-1. We particularly thank our INUIT partners for fruitful collaboration and sharing of ideas and IN samples. We also thank K. Dreischmeier and D. Cherian for testing and discussing features of the BINARY setup, and R. Herbert and B. Murray for helpful discussions on λ analysis.

Edited by: B. Ervens

References

Atkinson, J. D., Murray, B. J., Woodhouse, M. T., Whale, T. F., Baustian, K. J., Carslaw, K. S., Dobbie, S., O'Sullivan, D., and Malkin, T. L.: The importance of feldspar for ice nucleation by mineral dust in mixed-phase clouds, Nature, 498, 355–358, doi:10.1038/nature12278, 2013.

Attard, E., Yang, H., Delort, A.-M., Amato, P., Pöschl, U., Glaux, C., Koop, T., and Morris, C. E.: Effects of atmospheric conditions on ice nucleation activity of *Pseudomonas*, Atmos. Chem. Phys., 12, 10667–10677, doi:10.5194/acp-12-10667-2012, 2012.

Barahona, D.: On the ice nucleation spectrum, Atmos. Chem. Phys., 12, 3733–3752, doi:10.5194/acp-12-3733-2012, 2012.

Barlow, T. W. and Haymet, A. D. J.: ALTA: An automated lag-time apparatus for studying the nucleation of supercooled liquids, Rev. Sci. Instrum., 66, 2996, doi:10.1063/1.1145586, 1995.

Bigg, E. K.: The supercooling of water, Proc. Phys. Soc. Sect. B, 66, 688–694, doi:10.1088/0370-1301/66/8/309, 1953.

Cantrell, W. and Heymsfield, A.: Production of Ice in Tropospheric Clouds: A Review, B. Am. Meteorol. Soc., 86, 795–807, doi:10.1175/BAMS-86-6-795, 2005.

Connolly, P. J., Möhler, O., Field, P. R., Saathoff, H., Burgess, R., Choularton, T., and Gallagher, M.: Studies of heterogeneous freezing by three different desert dust samples, Atmos. Chem. Phys., 9, 2805–2824, doi:10.5194/acp-9-2805-2009, 2009.

Cziczo, D. J. and Froyd, K. D.: Sampling the composition of cirrus ice residuals, Atmos. Res., 142, 15–31, doi:10.1016/j.atmosres.2013.06.012, 2014.

Della Gatta, G., Richardson, M. J., Sarge, S. M., and Stølen, S.: Standards, calibration, and guidelines in microcalorimetry. Part 2. Calibration standards for differential scanning calorimetry (IUPAC Technical Report), Pure Appl. Chem., 78, 1455–1476, doi:10.1351/pac200678071455, 2006.

DeMott, P. J., Prenni, A. J., Liu, X., Kreidenweis, S. M., Petters, M. D., Twohy, C. H., Richardson, M. S., Eidhammer, T., and Rogers, D. C.: Predicting global atmospheric ice nuclei distributions and their impacts on climate, P. Natl. Acad. Sci. USA, 107, 11217–11222, doi:10.1073/pnas.0910818107, 2010.

Ervens, B. and Feingold, G.: On the representation of immersion and condensation freezing in cloud models using different nucleation schemes, Atmos. Chem. Phys., 12, 5807–5826, doi:10.5194/acp-12-5807-2012, 2012.

Ervens, B. and Feingold, G.: Sensitivities of immersion freezing: reconciling classical nucleation theory and deterministic expressions, Geophys. Res. Lett., 40, 3320–3324, doi:10.1002/grl.50580, 2013.

Fletcher, N. H.: Active sites and ice crystal nucleation, J. Atmos. Sci., 26, 1266–1271, doi:10.1175/1520-0469(1969)026<1266:ASAICN>2.0.CO;2, 1969.

Fornea, A. P., Brooks, S. D., Dooley, J. B., and Saha, A.: Heterogeneous freezing of ice on atmospheric aerosols containing ash, soot, and soil, J. Geophys. Res., 114, D13201, doi:10.1029/2009JD011958, 2009.

Hartmann, S., Augustin, S., Clauss, T., Wex, H., Šantl-Temkiv, T., Voigtländer, J., Niedermeier, D., and Stratmann, F.: Immersion freezing of ice nucleation active protein complexes, Atmos. Chem. Phys., 13, 5751–5766, doi:10.5194/acp-13-5751-2013, 2013.

Heneghan, A. F., Wilson, P. W., and Haymet, A. D. J.: Heterogeneous nucleation of supercooled water, and the effect of an added catalyst, P. Natl. Acad. Sci. USA, 99, 9631–9634, doi:10.1073/pnas.152253399, 2002.

Herbert, R. J., Murray, B. J., Whale, T. F., Dobbie, S. J., and Atkinson, J. D.: Representing time-dependent freezing behaviour in immersion mode ice nucleation, Atmos. Chem. Phys., 14, 8501–8520, doi:10.5194/acp-14-8501-2014, 2014.

Hiranuma, N., Möhler, O., Bingemer, H., Bundke, U., Cziczo, D. J., Danielczok, A., Ebert, M., Garimella, S., Hoffmann, N., Höhler, K., Kanji, Z. A., Kiselev, A., Raddatz, M., and Stetzer, O.: Immersion freezing of clay minerals and bacterial ice nuclei, in: Nucleation Atmos. Aerosols (AIP Conf. Proc. 1527), edited by: DeMott, P. J. and O'Dowd, C. D., 914–917, doi:10.1063/1.4803420, AIP Publishing, Melville, NY, USA, 2013.

Hoffer, T. E.: A laboratory investigation of droplet freezing, J. Meteorol., 18, 766–778, doi:10.1175/1520-0469(1961)018<0766:ALIODF>2.0.CO;2, 1961.

Hoose, C. and Möhler, O.: Heterogeneous ice nucleation on atmospheric aerosols: a review of results from laboratory experiments, Atmos. Chem. Phys., 12, 9817–9854, doi:10.5194/acp-12-9817-2012, 2012.

Hoyle, C. R., Pinti, V., Welti, A., Zobrist, B., Marcolli, C., Luo, B., Höskuldsson, Á., Mattsson, H. B., Stetzer, O., Thorsteinsson, T., Larsen, G., and Peter, T.: Ice nucleation properties of volcanic ash from Eyjafjallajökull, Atmos. Chem. Phys., 11, 9911–9926, doi:10.5194/acp-11-9911-2011, 2011.

Iannone, R., Chernoff, D. I., Pringle, A., Martin, S. T., and Bertram, A. K.: The ice nucleation ability of one of the most abundant types of fungal spores found in the atmosphere, Atmos. Chem. Phys., 11, 1191–1201, doi:10.5194/acp-11-1191-2011, 2011.

Jung, S., Tiwari, M. K., and Poulikakos, D.: Frost halos from supercooled water droplets, P. Natl. Acad. Sci. USA, 109, 16073–16078, doi:10.1073/pnas.1206121109, 2012.

Knopf, D. A. and Lopez, M. D.: Homogeneous ice freezing temperatures and ice nucleation rates of aqueous ammonium sulfate and aqueous levoglucosan particles for relevant atmospheric conditions, Phys. Chem. Chem. Phys., 11, 8056–8068, doi:10.1039/b903750k, 2009.

Knopf, D. A. and Alpert, P. A.: A water activity based model of heterogeneous ice nucleation kinetics for freezing of water and aqueous solution droplets, Faraday Discuss., 165, 513–534, doi:10.1039/c3fd00035d, 2013.

Koop, T. and Zobrist, B.: Parameterizations for ice nucleation in biological and atmospheric systems, Phys. Chem. Chem. Phys., 11, 10839–10850, doi:10.1039/b914289d, 2009.

Koop, T., Luo, B., Biermann, U. M., Crutzen, P. J., and Peter, T.: Freezing of HNO_3/H_2SO_4/H_2O solutions at stratospheric temperatures: nucleation statistics and experiments, J. Phys. Chem. A, 101, 1117–1133, doi:10.1021/jp9626531, 1997.

Koop, T., Ng, H. P., Molina, L. T., and Molina, M. J.: A new optical technique to study aerosol phase transitions: the nucleation of ice from H_2SO_4 aerosols, J. Phys. Chem. A, 102, 8924–8931, doi:10.1021/jp9828078, 1998.

Koop, T., Luo, B., Tsias, A., and Peter, T.: Water activity as the determinant for homogeneous ice nucleation in aqueous solutions, Nature, 406, 611–614, doi:10.1038/35020537, 2000.

Kulkarni, G., Fan, J., Comstock, J. M., Liu, X., and Ovchinnikov, M.: Laboratory measurements and model sensitivity studies of dust deposition ice nucleation, Atmos. Chem. Phys., 12, 7295–7308, doi:10.5194/acp-12-7295-2012, 2012.

Lamb, D. and Verlinde, J.: Physics and Chemistry of Clouds, Cambridge University Press, Cambridge, doi:10.1017/CBO9780511976377, 2011.

Linstrom, P. J. and Mallard, W. G. (Eds.): NIST Chemistry WebBook, NIST Standard Reference Database Number 69, National Institute of Standards and Technology, Gaithersburg MD, 20899, available at: http://webbook.nist.gov, last access: 8 September 2014.

Lohmann, U. and Diehl, K.: Sensitivity studies of the importance of dust ice nuclei for the indirect aerosol effect on stratiform mixed-phase clouds, J. Atmos. Sci., 63, 968–982, doi:10.1175/JAS3662.1, 2006.

Lüönd, F., Stetzer, O., Welti, A., and Lohmann, U.: Experimental study on the ice nucleation ability of size-selected kaolinite particles in the immersion mode, J. Geophys. Res., 115, D14201, doi:10.1029/2009JD012959, 2010.

Maki, L. R., Galyan, E. L., Chang-Chien, M. M., and Caldwell, D. R.: Ice nucleation induced by pseudomonas syringae, Appl. Microbiol., 28, 456–459, 1974.

Messerly, J. F., Guthrie, G. B., Todd, S. S., and Finke, H. L.: Low-temperature thermal data for pentane, n-heptadecane, and n-octadecane. Revised thermodynamic functions for the n-alkanes, C5–C18, J. Chem. Eng. Data, 12, 338–346, doi:10.1021/je60034a014, 1967.

Möhler, O., Georgakopoulos, D. G., Morris, C. E., Benz, S., Ebert, V., Hunsmann, S., Saathoff, H., Schnaiter, M., and Wagner, R.: Heterogeneous ice nucleation activity of bacteria: new laboratory experiments at simulated cloud conditions, Biogeosciences, 5, 1425–1435, doi:10.5194/bg-5-1425-2008, 2008.

Mondieig, D., Rajabalee, F., Metivaud, V., Oonk, H. A. J., and Cuevas-Diarte, M. A.: n-alkane binary molecular alloys, Chem. Mater., 16, 786–798, doi:10.1021/cm031169p, 2004.

Murphy, D. M. and Koop, T.: Review of the vapour pressures of ice and supercooled water for atmospheric applications, Q. J. Roy. Meteorol. Soc., 131, 1539–1565, doi:10.1256/qj.04.94, 2005.

Murray, B. J., Broadley, S. L., Wilson, T. W., Bull, S. J., Wills, R. H., Christenson, H. K., and Murray, E. J.: Kinetics of the homogeneous freezing of water, Phys. Chem. Chem. Phys., 12, 10380–10387, doi:10.1039/c003297b, 2010.

Murray, B. J., Broadley, S. L., Wilson, T. W., Atkinson, J. D., and Wills, R. H.: Heterogeneous freezing of water droplets containing kaolinite particles, Atmos. Chem. Phys., 11, 4191–4207, doi:10.5194/acp-11-4191-2011, 2011.

Murray, B. J., O'Sullivan, D., Atkinson, J. D., and Webb, M. E.: Ice nucleation by particles immersed in supercooled cloud droplets, Chem. Soc. Rev., 41, 6519–6554, doi:10.1039/c2cs35200a, 2012.

Niedermeier, D., Hartmann, S., Shaw, R. A., Covert, D., Mentel, T. F., Schneider, J., Poulain, L., Reitz, P., Spindler, C., Clauss, T., Kiselev, A., Hallbauer, E., Wex, H., Mildenberger, K., and Stratmann, F.: Heterogeneous freezing of droplets with immersed mineral dust particles – measurements and parameterization, Atmos. Chem. Phys., 10, 3601–3614, doi:10.5194/acp-10-3601-2010, 2010.

Niemand, M., Möhler, O., Vogel, B., Vogel, H., Hoose, C., Connolly, P., Klein, H., Bingemer, H., DeMott, P., Skrotzki, J., and Leisner, T.: A particle-surface-area-based parameterization of immersion freezing on desert dust particles, J. Atmos. Sci., 69, 3077–3092, doi:10.1175/JAS-D-11-0249.1, 2012.

O'Sullivan, D., Murray, B. J., Malkin, T. L., Whale, T. F., Umo, N. S., Atkinson, J. D., Price, H. C., Baustian, K. J., Browse, J., and Webb, M. E.: Ice nucleation by fertile soil dusts: relative importance of mineral and biogenic components, Atmos. Chem. Phys., 14, 1853–1867, doi:10.5194/acp-14-1853-2014, 2014.

Parks, G. S. and Huffman, H. M.: Some fusion and transition data for hydrocarbons, Ind. Eng. Chem., 23, 1138–1139, doi:10.1021/ie50262a018, 1931.

Pruppacher, H. R. and Klett, J. D.: Microphysics of Clouds and Precipitation, 2nd Edn., Kluwer Academic Publishers, New York, 1997.

Pummer, B. G., Bauer, H., Bernardi, J., Bleicher, S., and Grothe, H.: Suspendable macromolecules are responsible for ice nucleation activity of birch and conifer pollen, Atmos. Chem. Phys., 12, 2541–2550, doi:10.5194/acp-12-2541-2012, 2012.

Remmers, M. L.: Kerzenruß als Vorläufer für robuste und transparente superamphiphobe Beschichtungen, Ba thesis, Bielefeld University, Bielefeld, 2012.

Riechers, B., Wittbracht, F., Hütten, A., and Koop, T.: The homogeneous ice nucleation rate of water droplets produced in a microfluidic device and the role of temperature uncertainty, Phys. Chem. Chem. Phys., 15, 5873–5887, doi:10.1039/c3cp42437e, 2013.

Salcedo, D., Molina, L. T., and Molina, M. J.: Nucleation rates of nitric acid dihydrate in $1 : 2$ HNO_3/H_2O solutions at stratospheric temperatures, Geophys. Res. Lett., 27, 193, doi:10.1029/1999GL010991, 2000.

Sarge, S. M., Höhne, G. W., Cammenga, H. K., Eysel, W., and Gmelin, E.: Temperature, heat and heat flow rate calibration of scanning calorimeters in the cooling mode, Thermochim. Acta, 361, 1–20, doi:10.1016/S0040-6031(00)00543-8, 2000.

Steinke, I., Möhler, O., Kiselev, A., Niemand, M., Saathoff, H., Schnaiter, M., Skrotzki, J., Hoose, C., and Leisner, T.: Ice nucleation properties of fine ash particles from the Eyjafjallajökull eruption in April 2010, Atmos. Chem. Phys., 11, 12945–12958, doi:10.5194/acp-11-12945-2011, 2011.

Stopelli, E., Conen, F., Zimmermann, L., Alewell, C., and Morris, C. E.: Freezing nucleation apparatus puts new slant on study of biological ice nucleators in precipitation, Atmos. Meas. Tech., 7, 129–134, doi:10.5194/amt-7-129-2014, 2014.

Turner, M. A., Arellano, F., and Kozloff, L. M.: Three separate classes of bacterial ice nucleation structures, J. Bacteriol., 172, 2521–2526, 1990.

Vali, G.: Supercooling of water and nucleation of ice (drop freezer), Am. J. Phys., 39, 1125–1128, doi:10.1119/1.1976585, 1971a.

Vali, G.: Quantitative Evaluation of Experimental Results an the Heterogeneous Freezing Nucleation of Supercooled Liquids, J. Atmos. Sci., 28, 402–409, doi:10.1175/1520-0469(1971)028<0402:QEOERA>2.0.CO;2, 1971b.

Vali, G.: Freezing rate due to heterogeneous nucleation, J. Atmos. Sci., 51, 1843–1856, doi:10.1175/1520-0469(1994)051<1843:FRDTHN>2.0.CO;2, 1994.

Vali, G.: Repeatability and randomness in heterogeneous freezing nucleation, Atmos. Chem. Phys., 8, 5017–5031, doi:10.5194/acp-8-5017-2008, 2008.

Vali, G.: Interpretation of freezing nucleation experiments: singular and stochastic; sites and surfaces, Atmos. Chem. Phys., 14, 5271–5294, doi:10.5194/acp-14-5271-2014, 2014.

Vali, G. and Stansbury, E. J.: Time-dependet characteristics of the heterogeneous nucleation of ice, Can. J. Phys., 44, 477–502, doi:10.1139/p66-044, 1966.

Vali, G. and Snider, J. R.: Time-dependent freezing rate parcel model, Atmos. Chem. Phys. Discuss., 14, 29305–29329, doi:10.5194/acpd-14-29305-2014, 2014.

Vali, G., Christensen, M., Fresh, R. W., Galyan, E. L., Maki, L. R., and Schnell, R. C.: Biogenic Ice Nuclei. Part II: Bacterial Sources, J. Atmos. Sci., 33, 1565–1570, doi:10.1175/1520-0469(1976)033<1565:BINPIB>2.0.CO;2, 1976.

Vonnegut, B. and Baldwin, M.: Repeated nucleation of a supercooled water sample that contains silver iodide particles, J. Clim. Appl. Meteorol., 23, 486–490, doi:10.1175/1520-0450(1984)023<0486:RNOASW>2.0.CO;2, 1984.

Ward, P. J. and DeMott, P. J.: Preliminary experimental evaluation of Snomax Snow Inducer, Pseudomonas syringae, as an artificial ice nucleus for weather modification, J. Weather Modif., 21, 9–13, 1989.

Welti, A., Lüönd, F., Kanji, Z. A., Stetzer, O., and Lohmann, U.: Time dependence of immersion freezing: an experimental study on size selected kaolinite particles, Atmos. Chem. Phys., 12, 9893–9907, doi:10.5194/acp-12-9893-2012, 2012.

Welz, T.: Untersuchung der Eisnukleation in wässrigen Birkenpollensuspensionen, Ba thesis, Bielefeld University, Bielefeld, 2013.

Wex, H., Augustin-Bauditz, S., Boose, Y., Budke, C., Curtius, J., Diehl, K., Dreyer, A., Frank, F., Hartmann, S., Hiranuma, N., Jantsch, E., Kanji, Z. A., Kiselev, A., Koop, T., Möhler, O., Niedermeier, D., Nillius, B., Rösch, M., Rose, D., Schmidt, C., Steinke, I., and Stratmann, F.: Intercomparing different devices for the investigation of ice nucleating particles using Snomax® as test substance, Atmos. Chem. Phys., 15, 1463–1485, doi:10.5194/acp-15-1463-2015, 2015.

Wood, S. E., Baker, M. B., and Swanson, B. D.: Instrument for studies of homogeneous and heterogeneous ice nucleation in free-falling supercooled water droplets, Rev. Sci. Instrum., 73, 3988, doi:10.1063/1.1511796, 2002.

Wright, T. P. and Petters, M. D.: The role of time in heterogeneous freezing nucleation, J. Geophys. Res. Atmos., 118, 3731–3743, doi:10.1002/jgrd.50365, 2013.

Yankofsky, S. A., Levin, Z., Bertold, T., and Sandlerman, N.: Some basic characteristics of bacterial freezing nuclei, J. Appl. Meteorol., 20, 1013–1019, doi:10.1175/1520-0450(1981)020<1013:SBCOBF>2.0.CO;2, 1981.

Zobrist, B., Koop, T., Luo, B., Marcolli, C., and Peter, T.: Heterogeneous ice nucleation rate coefficient of water droplets coated by a nonadecanol monolayer, J. Phys. Chem. C, 111, 2149–2155, doi:10.1021/jp066080w, 2007.

8

ECOC comparison exercise with identical thermal protocols after temperature offset correction – instrument diagnostics by in-depth evaluation of operational parameters

P. Panteliadis[1], T. Hafkenscheid[2], B. Cary[3], E. Diapouli[4], A. Fischer[5], O. Favez[6], P. Quincey[7], M. Viana[8], R. Hitzenberger[9], R. Vecchi[10], D. Saraga[11], J. Sciare[12], J. L. Jaffrezo[13], A. John[14], J. Schwarz[15], M. Giannoni[16], J. Novak[17], A. Karanasiou[8], P. Fermo[18], and W. Maenhaut[19]

[1]Municipal Health Service (GGD) Amsterdam, Department of Air Quality, Amsterdam, the Netherlands
[2]National Institute for Public Health and the Environment, Bilthoven, the Netherlands
[3]Sunset Laboratory Inc, Tigard, Oregon, USA
[4]National Center for Scientific Research "Demokritos", Institute of Nuclear & Radiological Sciences & Technology, Energy & Safety, Athens, Greece
[5]EMPA – Swiss Federal Laboratories for Materials Science and Technology, Duebendorf, Switzerland
[6]INERIS, Verneuil-en-Halatte, France
[7]National Physical Laboratory, Teddington, UK
[8]Institute for Environmental Assessment and Water Research (IDAEA-CSIC), Barcelona, Spain
[9]Aerosolphysics and Environmental Physics, Faculty of Physics, Vienna, Austria
[10]Department of Physics, Università degli Studi di Milano, Milan, Italy
[11]Demokritos, National Center for Scientific Research, Environmental Research Laboratory, Athens, Greece
[12]Laboratoire des Sciences du Climat et de l'Environnement (LSCE), CEA-CNRS-UVSQ, Gif-sur-Yvette, France
[13]Univ. Grenoble Alpes, CNRS, LGGE, 38000 Grenoble, France
[14]Institute for Energy and Environmental Technology e.V. Air Quality & Sustainable Nanotechnology Division, Duisburg, Germany
[15]Institute of Chemical Process Fundamentals AS CR, Prague, Czech Republic
[16]Istituto Nazionale di Fisica Nucleare (INFN), Sezione di Firenze, Florence, Italy
[17]Czech Hydrometeorological Institute, Prague, Czech Republic
[18]Department of Chemistry, Università degli Studi di Milano, Milan, Italy
[19]Department of Analytical Chemistry, Ghent University, Gent, 9000, Belgium

Correspondence to: P. Panteliadis (ppanteliadis@ggd.amsterdam.nl)

Abstract. A comparison exercise on thermal-optical elemental carbon/organic carbon (ECOC) analysers was carried out among 17 European laboratories. Contrary to previous comparison exercises, the 17 participants made use of an identical instrument set-up, after correcting for temperature offsets with the application of a recently developed temperature calibration kit (Sunset Laboratory Inc, OR, US). Temperature offsets reported by participants ranged from -93 to $+100\,^\circ$C per temperature step. Five filter samples and two su-crose solutions were analysed with both the EUSAAR2 and NIOSH870 thermal protocols.

z scores were calculated for total carbon (TC); nine outliers and three stragglers were identified. Three outliers and eight stragglers were found for EC. Overall, the participants provided results between the warning levels with the exception of two laboratories that showed poor performance, the causes of which were identified and corrected through the course of the comparison exercise. The TC repeatability and reproducibility (expressed as relative standard deviations) were 11 and 15 % for EUSAAR2 and 9.2 and 12 % for NIOSH870; the standard deviations for EC were 15 and 20 % for EUSAAR2 and 20 and 26 % for NIOSH870.

TC was in good agreement between the two protocols, $TC_{NIOSH870} = 0.98 \times TC_{EUSAAR2}$ ($R^2 = 1.00$, robust means). Transmittance (TOT) calculated EC for NIOSH870 was found to be 20 % lower than for EUSAAR2, $EC_{NIOSH870} = 0.80 \times EC_{EUSAAR2}$ ($R^2 = 0.96$, robust means). The thermograms and laser signal values were compared and similar peak patterns were observed per sample and protocol for most participants. Notable deviations from the typical patterns indicated either the absence or inaccurate application of the temperature calibration procedure and/or pre-oxidation during the inert phase of the analysis. Low or zero pyrolytic organic carbon (POC), as reported by a few participants, is suggested as an indicator of an instrument-specific pre-oxidation. A sample-specific pre-oxidation effect was observed for filter G, for all participants and both thermal protocols, indicating the presence of oxygen donors on the suspended particulate matter. POC (TOT) levels were lower for NIOSH870 than for EUSAAR2, which is related to the heating profile differences of the two thermal protocols.

---

1   Introduction

Carbon in suspended atmospheric particulate matter usually falls into one of three wide categories, elemental carbon (EC), organic carbon (OC), and carbonate carbon (CC), the major fraction of inorganic carbon. Recently, more attention has been drawn to EC, due to its linkage to adverse health (Highwood and Kinnersley, 2006; Adar and Kaufman, 2007; Janssen et al., 2011, 2012) and climate effects (Jacobson, 2001; IPCC, 2007; Ramanathan and Carmichael, 2008). Several studies suggest that EC is a valid indicator for traffic emissions and include its analysis during monitoring campaigns (Lena et al., 2002; Schauer, 2003; Qadir et al., 2013; Panteliadis et al., 2014). A number of EC measurement techniques exist (Watson et al., 2005; Hitzenberger et al., 2006) with the thermal-optical transmittance (TOT) or reflectance (TOR) methods being broadly used in Europe and the USA. Several thermal protocols, which apply to TOT or TOR analysers, have been developed with NIOSH5040 (Birch and Cary, 1996), IMPROVE A (Chow et al., 2007) and EUSAAR2 (Cavalli et al., 2010) being the most commonly applied.

Even though quality assurance and quality control (QA/QC) procedures are of importance for any air quality measurements, no standard has yet been established in Europe for ECOC analysis (Chow et al., 2011). Following the EU Directive 2008/50/EC on ambient air and cleaner air for Europe, a technical report has been published (CEN TR 16243, 2011) and further work is currently being carried out towards method standardization within CEN-TC 264 (European Committee for Standardization) Working Group 35. Alongside the implementation of the technical report recommendations in the standard operation procedures of laboratories, comparison exercises are an important additional step towards QA/QC optimization.

The department of Air Quality of Public Health Service Amsterdam has been organizing laboratory comparison exercises for the past few years on thermal-optical ECOC analysers (Panteliadis, 2009a, 2011). To our knowledge, previous laboratory comparisons performed in Europe up to 2013 considered results derived from different protocols applied by each participant, usually NIOSH-like (CEN TR 16243, 2011) or EUSAAR2, on filter samples, limiting comparability of the performance of each laboratory. Numerous studies have demonstrated that ECOC analysis of ambient samples is sensitive to the temperature protocol selected (Sciare et al., 2003; Schauer et al., 2003; Chow et al., 2004; Cavalli et al., 2010; Zhi et al., 2011; Piazzalunga et al., 2011; Cheng et al., 2011, 2012). As a result, the temperature protocol selection may affect the conclusions obtained from comparisons between thermal-optical and optical (black carbon) analysis (Schmid et al., 2001; Schauer et al., 2003; ten Brink et al., 2004). Differences also occur with regard to the optical method used for the charring carbon correction, transmittance or reflectance, with the latter usually resulting in greater EC concentrations (Chow et al., 2004; Panteliadis, 2009b; Maenhaut et al., 2011).

The scope of this 2012 comparison exercise was to evaluate results based on an identical set-up for all participants using a lab ECOC analyser (Sunset Laboratory Inc., Tigard, OR, US). Generally in such exercises the same thermal protocol would be applied by all participants. However, the debate over the NIOSH-like and the EUSAAR2 thermal protocols is still ongoing in Europe, and the selection of a single temperature protocol would have been controversial and limiting. Comparison exercises performed so far have let the participants decide on the protocol applied (Panteliadis, 2009a, 2011; Emblico et al., 2012; Cavalli et al., 2012). As an alternative, we decided to stipulate the use of both NIOSH870, one of the latest versions of NIOSH-like protocols, and EUSAAR2 by each participant, providing additional information that could point out possible differences between the two protocols.

Each thermal protocol involves several temperature steps, and instrument-specific deviations from the desired temperature may alter the sample treatment and affect the analysis result (Chow et al., 2005). These deviations may originate from differences in type, age or installation of the heating coils used in each instrument. Since the introduction of a temperature calibration kit by the analyser's manufacturer (Sunset Laboratory Inc, OR, US) in early 2012, it has become possible to minimize such deviations. All participants using the lab ECOC analyser performed the calibration procedure and compensated for the temperature offsets before the comparison exercise analysis.

A common practice for total carbon (TC) calibrations and routine checks is the analysis of standard sucrose solutions. Such sucrose solutions were included in the current exercise in order to provide an insight into the degree of repeatability of these procedures, as well as to evaluate the practicability of adding a known volume of sucrose on the filter to be analysed. Unfortunately, the standard sucrose solutions can only provide information on calibration with pure OC. A suitable reference material consisting of pure EC is still lacking (Baumgardner et al., 2012). Finally, on top of the standard statistical analysis usually performed in such comparison exercises, a more in-depth evaluation of instrument specific parameters and characteristics, including the laser and the flame ionization detector (FID) signal as well as the peak distribution and calibration peak area, was carried out.

2 Methods

2.1 Sample preparation and distribution

A total of five 24 h PM loaded samples were collected for the exercise. Filters were selected with the intention of covering a common range of characteristics that occur in samples used for ECOC analysis. It is known, however, that the limited number of filters selected cannot be fully representative of the wide variety of ambient samples, which can be influenced by a number of parameters such as particle composition, pollution sources, seasonal and spatial variation.

The urban background sample from Amsterdam, The Netherlands, was collected with a PM_{10} high volume sampler (HVS) (ESM Andersen Instruments GmbH, Germany) on a Whatman QMA rectangular filter, 20.3×25.4 cm. The same filter type was used for the urban background sample from Athens, Greece, collected with a PM_3 GS2312 BL HVS (Tisch Environmental, Ohio, US). Two $PM_{2.5}$ suburban samples were collected in Duebendorf, Switzerland, with the use of a DHA80 sampler (Digitel Elektronik AG, Switzerland) on 150 mm diameter Pallflex Tissuquartz filters, on two consecutive dates. The same type of sampler and filter was used for the urban sample collected in Berne, Switzerland. The amount of transmitted laser light compared to the blank value, per filter, was reported by the organizing laboratory. An overview of the filters characteristics and sampling details is presented in Table 1.

Upon receipt at GGD Amsterdam, all filters were stored at a temperature below $5\,°C$ till the distribution date. Four rectangular punches of 1×1.5 cm^2 per participant were cut out from different sections and increasing distance from the centre of each filter to avoid any sampling bias. For the same reason no punches were cut from the sampled area close to the edges of the filters. The punches were then stored in separate, closed, Petri-slide dishes, which were sent to each par-

ticipant together with 30 mL vials of two standard sucrose solutions, S1 and S2, with nominal OC concentrations of 10.0 and $33.6\,\mu g\,10\,\mu L^{-1}$, respectively. For the two participants using a field instrument four circular punches of $2\,cm^2$ were prepared instead.

The homogeneity of PM loaded HVS filters, for the exact same samplers and filter media, had already been tested by GGD Amsterdam, and resulted in relative standard deviations of 11 % for EC, 6 % for OC and 5 % for TC, for 150 mm Pallflex Tissuquartz HVS filters, and 10 % for EC, 9 % for OC and 6 % for TC for Whatman QMA HVS filters (Table S1 in the Supplement). These values, however, represent only an indication of the expected within sample standard deviation for filter samples used in the current comparison exercise.

2.2 ECOC analysis

The EC, OC and TC concentrations in the PM samples and sucrose solutions were determined by all participants with the use either of a lab ECOC aerosol analyser (15 participants), or a semi-continuous ECOC field analyser (2 participants), all manufactured by Sunset Laboratory Inc. (Tigard, OR, US).

During the analysis, OC desorbs from the quartz fibre filter through progressive heating under a pure He stream, while a fraction of OC chars and forms pyrolysed organic carbon (POC). The sample is then heated in temperature steps under a mixture of 98 % He − 2 % O_2 (HeOx phase) and the POC and EC are desorbed. In order to correct for the pyrolysis effect, the analyser utilizes a 658 nm laser beam, reflected and/or transmitted through the filter media. The split point, which separates OC from EC and compensates for POC, is determined when the laser signal returns to its initial value. OC, EC and POC are catalytically converted, initially to CO_2 and then to CH_4, which is quantified with the use of an FID. The time necessary for the gaseous compounds desorbed to reach the FID unit from the filter media is defined as transit time and is an instrument-specific parameter. A fixed volume of calibration gas (5 % CH_4 in helium) is injected in the instrument at the end of each analysis and the responding FID signal forms the calibration peak. The area of the calibration peak together with a calibration constant are used for the calculation of the sample concentration. The calibration constant depends on the calibration gas fixed volume analysed per run, which is set by the manufacturer, and is instrument-specific.

The operating parameters of the analyser vary with the thermal protocol used during analysis. The two protocols mainly used in Europe, NIOSH870 and EUSAAR2, were applied by all participants for the analysis of the sample punches provided. Before analysis, most participants calibrated their instruments for temperature offsets at each step, with the use of a calibration kit. The temperature calibration procedure has to be applied for each thermal protocol separately, since they vary in the number of steps, tempera-

Table 1. PM loaded filters description and sampling details.

Filter code	Location	Site description	PM fraction ($\mu g\,m^{-3}$)	Filter type	Date	Volume (m^3)	Sampling time (h)	Instrument	% Transmitted laser intensity compared to blank
A	Athens	Urban	3.0 (NA)[2]	Whatman QMA	1 Mar 2012	1411	24	Andersen GS2312	38
		background		20.3 × 25.4 cm				BL HVS	
B	Berne	Urban	2.5 (72.8)	Pallflex Tissuquartz 150 mm	9 Feb 2012	720	24	Digitel DHA80	14
D	Duebendorf	Suburban	2.5 (7.8)	Pallflex Tissuquartz 150 mm	15 Feb 2012	720	24	Digitel DHA80	51
G	Amsterdam	Urban	10 (24.4)	Whatman QMA	13 Dec 2005[1]	1625	24	Andersen/GMW	32
		background		20.3 × 25.4 cm				1200 HVS	
U	Duebendorf	Suburban	2.5 (37.0)	Pallflex Tissuquartz 150 mm	14 Feb 2012	720	24	Digitel DHA80	23

[1] Stored below 5 °C till the distribution date; [2] not available.

Table 2. Details of the two thermal protocols applied by the participants and observed temperature offset range per step.

Carrier gas	NIOSH870			EUSAAR2		
	Time (s)	Temperature (°C)	Range T offsets (°C)	Time (s)	Temperature (°C)	Range T offsets (°C)
Purge time	10	–	–	10	–	–
Helium	80	310	(−58–46)	120	200	(−71–100)
Helium	80	475	(−51–63)	150	300	(−67–54)
Helium	80	615	(−50–70)	180	450	(−60–47)
Helium	110	870	(−68–81)	180	650	(−58–51)
OC analysis time	360			640		
Helium (Oven cool)	45	550	(−93–65)	30	–	–
Oxygen in helium (2 %)	45	550	(−93–65)	120	500	(−68–49)
Oxygen in helium (2 %)	45	625	(−75–67)	120	550	(−59–40)
Oxygen in helium (2 %)	45	700	(−65–70)	70	700	(−68–51)
Oxygen in helium (2 %)	45	775	(−70–72)	80	850	(−85–64)
Oxygen in helium (2 %)	45	850	(−76–73)	–	–	–
Oxygen in helium (2 %)	110	870	(−80–63)	–	–	–
EC analysis time	380			390		
Calibration	120			110		
Total analysis time	14 min and 20 s			19 min and 30 s		

ture and duration. The differences between EUSAAR2 and NIOSH870, together with the observed offset ranges, are illustrated in Table 2.

Two participants performed the temperature calibration only for the EUSAAR2 protocol while one did not apply it at all. Two participants used the temperature offsets as found for the GGD instrument. The temperature calibration procedure was not applicable for the field analysers. A wide range of temperature offsets was observed between participants (−93 to +100 °C), and also between different temperature steps for the same analyser (up to 86 °C). An overview of the thermal protocols, optical method and temperature offsets applied by each participant can be found in Table 3. The heating profile of each analyser after the temperature offset correction was also recorded for both protocols (Figs. S1–S2 in the Supplement).

Two of the four punches received by each participant were analysed with NIOSH870 and two with EUSAAR2, whereas triplicate analyses with both protocols were applied for the two sucrose solutions.

Table 3. Thermal protocols used for replicate analysis, optical method applied and temperature offset range per laboratory.

Laboratory	Protocols – replicates	Optical method	T offset range (°C)
1	NIOSH870 – EUSAAR2	TOT – TOR	(−58−−10)
2	NIOSH870 – EUSAAR2	TOT – TOR	(−65−−6)
3	NIOSH930[a] – EUSAAR2	TOT – TOR	(−87–21)
4	NIOSH870 – EUSAAR2	TOT	(−86−−50)
5	NIOSH-like[b] – EUSAAR2	TOT – TOR	(−55−−5)
6	NIOSH870 – EUSAAR2	TOT – TOR	(−93−−7)
7	NIOSH870 – EUSAAR2[c]	TOT – TOR	(−60−−24)
8	NIOSH870 – EUSAAR2	TOT – TOR	(30–81)
9	NIOSH870 – EUSAAR2	TOT	(−73−−31)
10	NIOSH870 – EUSAAR2	TOT – TOR	(−12–3)[d]
11	NIOSH870 – EUSAAR2	TOT	(−64−−16)[e]
12	NIOSH870 – EUSAAR2	TOT	(−59−−26)
13	NIOSH870 – EUSAAR2	TOT – TOR	(33–100)[d]
14	NIOSH870 – EUSAAR2	TOT – TOR	(−58−−10)[d]
15	NIOSH870 – EUSAAR2	TOT	Not performed
16	NIOSH870 – EUSAAR2	TOT	Not applicable[g]
17	NIOSH870 – EUSAAR2[f]	TOT	Not applicable[g]

[a] NIOSH930 applies an additional heating step at the end of the HeOx phase at 930 °C for 120 s; [b] 890 °C applied at the last heating step instead of 870 °; [c] limited number of filters analysed in EUSAAR2; [d] applied only for EUSAAR2 protocol; [e] no calibration performed, GGD offsets applied instead; [f] limited sample set analysed; [g] field analyser.

2.3 Statistical analysis

2.3.1 Laboratory performance

The robust means for the filter samples, derived from the means of replicate measurements, were calculated following ISO 13528 (2005, Annex C). Due to the lack of a certified reference material, the robust mean for each sample was taken as the consensus reference value. The z scores were calculated for TC and EC, and for EUSAAR2 and NIOSH870, in order to evaluate the capacity of each laboratory to comply with the selected fit-for-purpose standard deviation, using the equation:

$$z = (x - X)/\sigma, \tag{1}$$

where x is the result of the participant (average of duplicate analyses), X is the robust mean, σ is the fit-for-purpose standard deviation.

Due to the lack of certified reference methods for ECOC measurements, the fit-for-purpose standard deviations were arbitrarily selected, based on experience and the desired level for compliance purposes: 8.3 % for TC and 25 % for EC. The 8.3 % for TC roughly corresponds to a range of ±25 % into which all results should fall. z scores between the warning levels, −2 and +2, were considered as indications of satisfactory performance, while z scores between the warning and the action levels, −3 and +3, were considered questionable. All z scores outside the action levels range were considered as indications of unsatisfactory performance.

2.3.2 Method performance

The laboratory performance was initially evaluated graphically, by sucrose solutions plots and Mandel's h and k statistics for TC and EC results pooled for both EUSAAR2 and NIOSH870. The Mandel's h statistic indicated the between-laboratory consistency while Mandel's k indicated the within-laboratory consistency. Laboratory results reported above the critical value at 1 % significance level were identified as possible outliers, and between the critical values of 1 and 5 % significance level as stragglers (ISO 5725-2, 1994). Grubbs' and Cochran's statistical outlier tests were also applied and outliers were removed from the data set for the calculations of the corrected robust means, repeatability and reproducibility relative standard deviations (ISO 5725-2, 1994).

3 Results

3.1 Data evaluation

All results, as reported by the participating laboratories, namely TC, EC, OC and EC/TC for both TOT and TOR are given in the Supplement, Tables S2–S8. The reported TC concentrations ranged on average from 10.1 to 79.0 μg cm^{-2}, while EC ranged from 0.9 to 11.5 μg cm^{-2} (TOT) and 1.8 to 17.5 μg cm^{-2} (TOR), depending on the thermal protocol used.

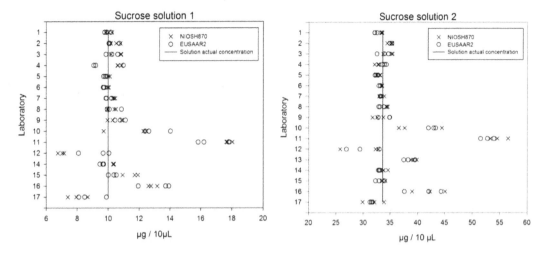

Figure 1. Triplicate sucrose solution analysis (S1 and S2) per protocol and participant. S1 and S2 concentrations of 10.0 and 33.6 μg 10 μL^{-1}, respectively.

3.2 Laboratory performance

An initial overview of deviation in performance can be gained from Fig. 1, which presents graphically the reported results of all participants for the two sucrose solutions.

The z scores for the EC and TC results of the filter samples, calculated separately for EUSAAR2 and NIOSH870, are shown in Figs. S3–S6 in the Supplement. For TC, seven outliers and three stragglers were identified for EUSAAR2, and 12 outliers for NIOSH870, all of which were reported by three participants. For EC, one outlier and four stragglers were identified for EUSAAR2 while two outliers and four stragglers were identified for NIOSH870. All outliers and stragglers were reported by three participants, two of whom were the same as those with high TC z scores.

3.3 Method performance

Figure S7, in the Supplement, presents the Mandel's k statistic for the sucrose solutions, with the EUSAAR2 and NIOSH870 TC pooled results. Five outliers were identified, two for laboratory 10, two for 12, and one for 16. The outliers of laboratories 10 and 12 were confirmed by Cochran's test while the one of laboratory 16 was identified as a straggler. The Mandel's h statistic values for the sucrose solutions can be found in Fig. S8 in the Supplement. Two outliers were found for laboratory 11, confirmed also by Grubbs' test.

Figure S9, in the Supplement, presents the Mandel's k statistic values for the loaded filters, for EUSAAR2 and NIOSH870 pooled TC results. Seven outliers were identified, three for laboratory 12, two for 16, and one for each of 10 and 11. Two stragglers were also identified, one for laboratory 13, and one for 16. Two out of the three outliers of laboratory 12, both for 16 and the one for 10 were confirmed by Cochran's test while the one for laboratory 11 was identified as a straggler.

The Mandel's h statistic values for the loaded filters, EUSAAR2 and NIOSH870 pooled TC results, can be found in Fig. S10 in the Supplement. Five outliers were identified, all for laboratory 11, four of which were indicated as stragglers by Grubbs' test. Four stragglers were observed, all for laboratory 10, one of which was confirmed by Grubbs' test.

Similarly to Fig. S9, Fig. S11 presents the Mandel's k statistic values for pooled EC (TOT) results of the loaded filters. Three outliers and two stragglers were identified; one outlier for each of laboratories 3, 4 and 16, and one straggler for 15 and 16. The three outliers were also confirmed by Grubbs' test. Figure S12, in the Supplement, presents the Mandel's h statistic values for the filters pooled EC (TOT) results. Four outliers and three stragglers were identified in total. Laboratories 10 and 11 reported two outliers and one straggler each while one straggler was reported by laboratory 3 and one by 8. The Grubbs' test confirmed all outliers and stragglers for laboratories 10 and 11 but not the stragglers for 3 and 8.

The normalized mean values and the repeatability and reproducibility relative standard deviations for the filter samples were calculated initially for the full data set and then after discarding the verified outliers. Table S9 in the Supplement shows the values separated by protocol for TC, while Table S10, also in the Supplement, shows the same for EC. For the corrected results, the repeatability relative standard deviation for TC was 11 % for EUSAAR2 and 9 % for NIOSH870. The reproducibility standard deviation was 15 % for EUSAAR2 and 12 % for NIOSH870. For EC, the repeatability standard deviation was 15 % for EUSAAR2 and 20 % for NIOSH870. The reproducibility standard deviation was 20 % for EUSAAR2 and 26 % for NIOSH870. All standard deviations were higher for EC than TC. All standard deviation values were higher for EUSAAR2 for TC, while the opposite held for EC.

The TC robust means, repeatability and reproducibility standard deviations per filter were calculated for pooled EU-SAAR2 and NIOSH870 results (Table S11 in the Supplement). TC robust means ranged from 9.7 to $77.0\,\mu\mathrm{g\,cm}^{-2}$. Repeatability ranged from 9 to 12 % and reproducibility from 11 to 15 % per filter.

No significant differences were observed between EUSAAR2 and NIOSH870 for TC, where $\mathrm{TC_{NIOSH870}} = 0.97 \times \mathrm{TC_{EUSAAR2}}$ $(R^2 = 0.96)$ for loaded PM filters and $\mathrm{TC_{NIOSH870}} = 1.00 \times \mathrm{TC_{EUSAAR2}}$ $(R^2 = 0.98)$ for sucrose solution raw data (Fig. S13 in the Supplement). When the raw data for EC for the loaded filters were compared, EUSAAR2 was found to report higher values, $\mathrm{EC_{NIOSH870}} = 0.73 \times \mathrm{EC_{EUSAAR2}}$ $(R^2 = 0.72)$ for TOT. For TOR, EUSAAR2 and NIOSH870 were closer, $\mathrm{EC_{NIOSH870}} = 0.85 \times \mathrm{EC_{EUSAAR2}}$ $(R^2 = 0.69)$ (Fig. S14 in the Supplement). For both protocols the use of TOR resulted in notably higher EC concentrations than TOT, 64 % higher $(R^2 = 0.52)$ for EUSAAR2 and 113 % higher $(R^2 = 0.44)$ for NIOSH870 $(N = 89)$. All zero NIOSH870 EC concentrations shown in the graphs were reported by a single participant due to laser failure. Note that not all participants reported data for both TOT and TOR, as a result of instrument configuration.

When the robust means were used for the same plots $\mathrm{TC_{NIOSH870}} = 0.98 \times \mathrm{TC_{EUSAAR2}}$ $(R^2 = 1.00)$ for loaded PM filters and $\mathrm{EC_{NIOSH870}} = 0.80 \times \mathrm{EC_{EUSAAR2}}$ $(R^2 = 0.96)$ for TOT and $\mathrm{EC_{NIOSH870}} = 1.15 \times \mathrm{EC_{EUSAAR2}}$ $(R^2 = 0.95)$ for TOR were found (Fig. S15 in the Supplement).

3.4 EUSAAR2 and NIOSH870 comparison

Table 4 shows the ranges of split points reported by the participants for each filter sample, protocol and optical method used. In all cases the split points in EUSAAR2 occur later than in NIOSH870, due to the extended overall duration of EUSAAR2. The split point defined by TOR appears to occur earlier than that defined by TOT. In most cases the overall split point range among participants for the same filter is ~ 100 s, except for sample A, where it extends up to 200 s, and sample G, up to 300 s. Both samples A and G were collected on the same filter media, Whatman QMA, while B, D and U are on Pallflex Tissuquartz, which may relate to the observed deviation. Note that Whatman QMA is thicker, $450\,\mu\mathrm{m}$, than Pallflex Tissuquartz, $432\,\mu\mathrm{m}$, as reported by their manufacturers.

The relative standard deviations of the calibration area of each instrument together with the calibration constant, the transit time and the atmospheric pressure are presented in Table 5. The calibration area relative standard deviations range from 1.4 to 24.6 % and the calibration constant from 17.1 to 28.7. Fluctuations of the atmospheric pressure of the laboratory may be of influence to the gas flows and consequently to the calibration area. The fluctuations in atmospheric pressure, as measured by each instrument, were in most cases

Table 4. Split point ranges as reported by all participants, per sample, protocol and optical method.

Sample	Protocol	Split point range (s)	
		Transmittance	Reflectance
A	EUSAAR2	(745–941)	(772–862)
	NIOSH870	(434–531)	(344–518)
B	EUSAAR2	(826–932)	(772–886)
	NIOSH870	(427–518)	(414–499)
D	EUSAAR2	(875–997)	(840–982)[a]
	NIOSH870	(490–593)[b]	(467–569)
G	EUSAAR2	(582–859)	(736–888)
	NIOSH870	(302–524)	(324–521)
U	EUSAAR2	(808–977)	(809–898)[c]
	NIOSH870	(471–561)	(347–534)

One participant reported clearly outlying split points: [a] 1073 s; [b] 819 s; [c] 991 s.

within a range of 10 mmHg. Nevertheless, five instruments reported values significantly lower than expected, between 374 and 427 mmHg, indicating a possible malfunction or absence of the pressure sensor, which results in the instrument recording an offset value.

The transit time ranges from 6 to 15 s. An incorrect transit time will result in a shift of the split point and thus incorrect concentration values for the OC and EC fractions. A check procedure for the transit time is available from the analyser's manufacturer but its application was not included in the prerequisites of the current comparison exercise.

Thermograms for sample A from different participants using NIOSH870 showed a similar peak distribution with a high first peak, low second, third and fourth peaks and a high fifth peak when POC and EC evolve. The split point range covers almost the whole fifth peak and no EC is measured after 600 s (Fig. 2). The EUSAAR2 thermograms for sample A show more variation in the peak distribution than the NIOSH870 thermograms, but in general start with a high first peak, with the second, third, and fourth peaks gradually reducing in size. Then a high fifth peak is observed when the POC and EC desorb, followed by a gradual decrease in the sixth and seventh peaks. EC seems to desorb up to the very end of the analysis (Fig. 3). Similar observations can be made for filter B (Figs. S16–S17).

For the sucrose solutions, with the exception of a couple of participants, there is good agreement in the peak distribution, which is better for NIOSH870 than EUSAAR2 (Figs. 4, 5). For NIOSH870, most of the OC is evolved in the first peak, followed by low second, third and fourth peaks and a slightly higher fifth peak for the POC. Almost no carbon is measured after 600 s. For EUSAAR2, most of the participants reported a high second peak and lower third, fourth, and fifth peaks. The POC evolves in a wide time frame from 700 s until the end of the analysis in some cases. Four participants report a medium to low first peak, while for the rest of the partici-

Table 5. Relative standard deviations (% rSD) of calibration area for all analyses per participant. Instrument specific characteristics, calibration constant, transit time and atmospheric pressure.

Laboratory	% rSD calibration area	Calibration constant	Transit time (seconds)	Atmospheric pressure (mmHg)
1	1.9	23.4	11	(729.3–744.1)
2	2.0	19.9	11	(426.1–426.9)
3	14.9	24.7	8	(751.4–763.6)
4	1.4	17.1	7	(741.4–747.7)
5	2.8	24.6	13	(748.3–750.9)
6	3.0	24.6	12	(725.8–739.6)
7	2.3	20.9	15	(761.7–775.0)
8	1.4	25.5	7	(384.3–384.4)
9	14.4	23.4	6	(742.2–746.5)
10	14.6	20.1	12	(375.4–375.7)
11	4.6	28.7	12	(774.6–785.0)
12	24.6	18.6	11	(374.6–374.9)
13	8.9	22.0	14	(382.0–382.6)
14	10.2	22.3	13	(723.5–726.3)
15	3.8	22.5	7	(719.7–731.4)

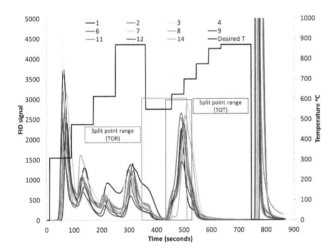

Figure 2. Thermograms of ECOC analysis on PM loaded quartz fibre filter (A sample), by NIOSH870 for all participants.

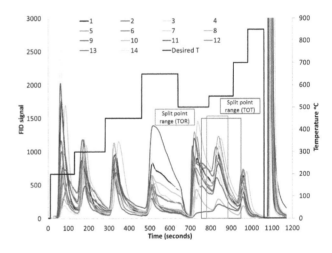

Figure 3. Thermograms of ECOC analysis on PM loaded quartz fibre filter (A sample), by EUSAAR2 for all participants.

pants it is not present. This first peak may be an indication of contamination due to handling during the sucrose analysis procedure. Such contamination would not be visible with the NIOSH870 protocol as it would be masked by the already high first peak.

Figures 6 and 7 show the concentrations for each temperature step and protocol for samples A and B. In most cases POC and EC are lower for NIOSH870 compared to EU-SAAR2. Some participants reported limited or no POC for sample A, suggesting that oxygen may have been present in the system during the inert phase. This became less visible for sample B, where concentrations were higher. For sucrose, most participants report the greatest fraction of OC at the first temperature step (OC1) for NIOSH870 and the second

(OC2) for EUSAAR2 (Fig. 8). Similarly to filter A, some participants report low or no POC for the sucrose solutions, indicating instrument specific pre-oxidation.

Figures 9–12 show the laser transmittance signal plots for each participant for the analysis of samples A and G for both EUSAAR2 and NIOSH870. A high frequency noise to the laser signal can be observed in all cases for laboratory 5, and non-systematic deviating behaviour for laboratories 4 and 11. Laboratory 12 shows a low frequency noise for all samples. A ramping up of the laser signal before the HeOx phase, indicating pre-oxidation, is seen occasionally for several laboratories, but mostly for 1, 3 and 4. The same ramping effect in the He phase, seen by almost all participants for filter G, indicates that in this case the pre-oxidation is not instrument spe-

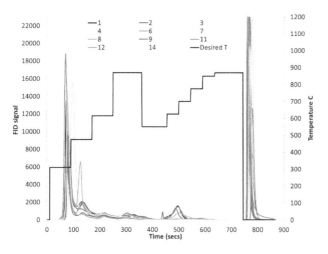

Figure 4. Thermograms of ECOC analysis on standard sucrose solution (S2), by NIOSH870 for all participants.

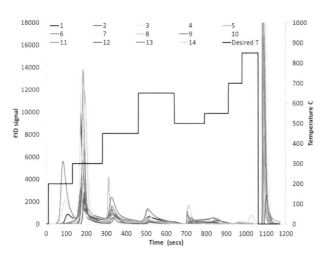

Figure 5. Thermograms of ECOC analysis on standard sucrose solution (S2), by EUSAAR2 for all participants.

cific but sample related. A possible explanation may be the presence of metals and/or metal oxides (Wang et al., 2010). The laser transmittance signal for a blank filter, as derived from the values at the end of each analysis, covers a wide range among the participants, from ~ 1000 to $\sim 20\,000$. The respective figures for the laser reflectance signal plots can be found in the Supplement, Figs. S18–S21. However, no observations similar to those for the transmittance signal can be drawn.

A limited number of participants reported data from instrument blank analysis. Supplement Figs. S22 and S23 show the laser signal, TOT and TOR results, for EUSAAR2 and NIOSH870 for the blank filters. The laser signal is stable throughout each analysis with the exception of laboratory 3, showing no dependence on the oven temperature.

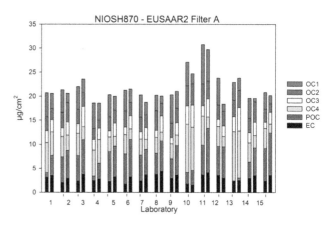

Figure 6. OC, POC and EC (TOT) concentrations (μg cm^{-2}) per temperature step and protocol for PM loaded quartz fibre filter (sample A).

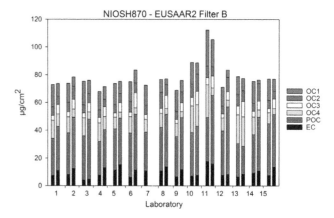

Figure 7. OC, POC and EC (TOT) concentrations (μg cm^{-2}) per temperature step and protocol for PM loaded quartz fibre filter (sample B).

4   Discussion

Based on z scores for TC, two laboratories showed poor performance, reporting results with a significant positive systematic error; seven outliers and two stragglers for laboratory 10 and ten outliers for laboratory 11 (Figs. S3, S4). In the course of this work the causes of the observed deviations were identified and corrected. Laboratory 16, which used a semi-automated field analyser, reported two outliers and one straggler, while the rest of participants were within the warning levels. Similar observations can be made for the EC z scores, with laboratory 10 reporting one outlier and three stragglers, laboratory 11 one outlier and four stragglers, and laboratory 3 one outlier and one straggler. It should be noted that different fit-for-purpose deviations were selected as levels of satisfactory performance for TC (8.3 %), which is thermally defined, and EC (25 %) which is both thermally and optically defined.

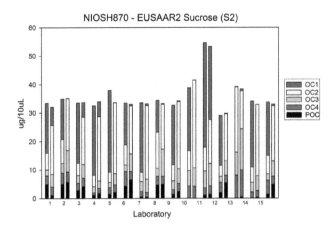

Figure 8. OC, POC and EC (TOT) concentrations (μg $10\,\mu\text{L}^{-1}$) per temperature step and protocol for sucrose solution (S2).

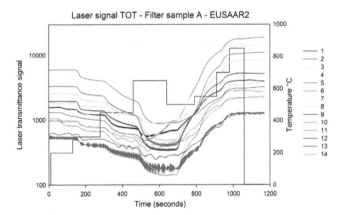

Figure 9. Laser transmittance signal during filter sample A analysis with the use of EUSAAR2 thermal protocol for all laboratories.

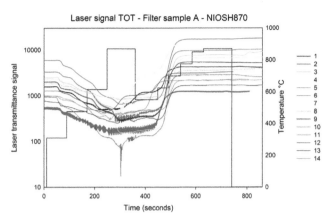

Figure 10. Laser transmittance signal during filter sample A analysis with the use of NIOSH870 thermal protocol for all laboratories.

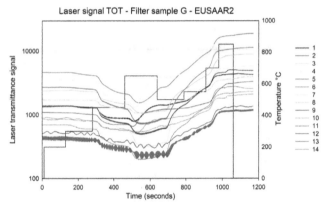

Figure 11. Laser transmittance signal during filter sample G analysis with the use of EUSAAR2 thermal protocol for all laboratories.

As seen in Fig. 1, the analyses of sucrose solutions in comparison exercises can serve as an indicator of poorly conducted TC calibration or unsatisfactory performance. Furthermore, it is clear that the repeatability derived from the analysis of a known volume of sucrose solution is dependent on the laboratory while independent of the thermal protocol used. When performed properly, analysis of standard sucrose solutions can provide a reliable TC calibration procedure. Only three participants showed problems in the within-laboratory consistency, and only one in the between-laboratory consistency (Figs. S7, S8).

No clear differences were noticeable between the z scores for EUSAAR2 and NIOSH870, suggesting that poor laboratory performance or deviating results are not protocol specific. The TC repeatability and reproducibility relative standard deviations, 12 and 15 % for EUSAAR2 and 9 and 12 % for NIOSH870, respectively, were at satisfactory levels, taking into account the homogeneity of similar PM sampled HVS filters that ranged from 6 to 10 %. All TC standard deviations were found lower for NIOSH870, which may be ex-

plained by the fact that all TC evolved relatively early during analysis while a larger fraction remained present in later steps for EUSAAR2. It may have been the case that a small fraction of TC did not evolve with EUSAAR2 for some samples, possibly those more highly loaded, resulting in greater standard deviations. Nevertheless, the findings of the current exercise showed that after the temperature calibration, almost identical concentrations of TC were measured by both protocols, $\text{TC}_{\text{NIOSH870}} = 0.98 \times \text{TC}_{\text{EUSAAR2}}$ ($R^2 = 1.00$) for robust means.

The EC repeatability and reproducibility relative standard deviations, 15 and 20 % for EUSAAR2 and 20 and 26 % for NIOSH870, respectively, were greater than the TC ones, indicating the additional uncertainties associated with the optical determination of EC. This is supported by the wide range of split points that varied by more than 200 s for the same sample among different participants. The EC standard deviations were higher for NIOSH870, probably due to the fact that the split point was located on high sections of FID peaks so deviations of a few seconds would have resulted in relatively large changes in the EC amount reported. For EU-

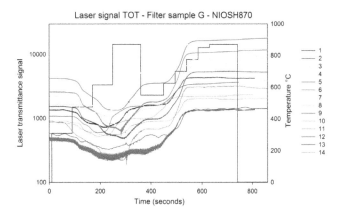

Figure 12. Laser transmittance signal during filter sample G analysis with the use of NIOSH870 thermal protocol for all laboratories.

SAAR2, the split point usually fell in less high sections of FID peaks. Furthermore, the wide range of the laser transmittance signal value for a blank filter, from ~ 1000 to $\sim 20\,000$, may also have affected the capacity of specific instruments to determine the split point accurately. Pre-oxidation resulting in early desorption of POC and EC could also have altered the split point's position.

Based on robust means, TOT EC reported by NIOSH870 is 20 % lower than by EUSAAR2, $EC_{NIOSH870,TOT} = 0.80 \times EC_{EUSAAR2,TOT}$ ($R^2 = 0.96$). Similar results were found in a comparison performed by GGD Amsterdam, prior to temperature calibration, with $EC_{NIOSH870,TOT} = 0.83 \times EC_{EUSAAR2,TOT}$ ($R^2 = 0.94$) (Fig. S24 in the Supplement). The above suggests that the temperature offset corrections resulted in no particular improvement in the agreement between thermal protocols. However, the selection of a thermal protocol clearly has an influence on EC analysis and should therefore be stated whenever results are reported, for clarity and comparability.

An additional parameter influencing the EC results is the optical method used. As expected, TOR results were higher than TOT for EC, 64 % higher for EUSAAR2 ($R^2 = 0.52$) and 113 % higher for NIOSH870 ($R^2 = 0.44$). However, the above values are based on raw data obtained by a limited number of participants that reported both TOR and TOT results, 10 out of a total of 17. The EC data from the different protocols also varied when applying TOR but in the opposite sense to TOT. More specifically, TOR EC measured with NIOSH870 resulted in higher values by 15 % compared to EUSAAR2, $EC_{NIOSH870,TOR} = 1.15 \times EC_{EUSAAR2,TOR}$ ($R^2 = 0.95$).

The stability of the calibration area throughout sequential analysis could serve as an additional indicator of robustness of the instrument, since observed deviations are usually caused by gas flow fluctuations, leakages or oven failures. A typical relative standard deviation (% rSD) of the calibration areas reported throughout an analysis run would be below 5 %. Nevertheless, in some cases it was reported to be much

higher, mostly due to the fact that analysis was performed on different analysis days (Table 5). In that case, gas flows were shut down at the end of one day and re-adjusted on the following one. It should be noted that gas flows are not strictly defined but have to lie within a suggested range, set manually by the user. The calibration area can also be affected when the calibration gas cylinder is replaced. While the nominal concentration of CH_4 in helium is 5 %, deviations are possible. A verification of the concentration of the calibration gas is recommended using an independent standard such as a sucrose solution, every time a cylinder is replaced.

When the thermograms were compared across participants, patterns in peak distribution were seen for each sample and protocol used. Overlapping peaks were observed for both protocols but these were not investigated in detail since the focus was mainly given to the EC and OC fractions and not the separate temperature step fractions. Peaks evolving systematically earlier or later may relate to an error in the determination of the transit time of the instrument (Figs. 2–5). Variations in peak distribution may be the result of absent or inaccurate application of the temperature calibration procedure, or even an indication of possible contamination.

Pre-oxidation is another potential cause of peak distribution differences, which can be verified by the POC concentrations. Low or zero reported POC during analysis may be an indication of oxygen entering the main oven during analysis in the helium phase (Figs. 6–8). Pre-oxidation was identified more clearly during sucrose solution analysis, when peaks evolving in the oxygen phase were small or non-existent and no or low POC was measured (Figs. 4, 5, 8). A possible cause of pre-oxidation may be traces of oxygen present in the helium stream. These can be eliminated by using an in-line oxygen trap.

Leakages, loose connections and oxygen in the helium stream are potential causes of pre-oxidation that may result in a constant presence of oxygen in the instrument during the inert phase of analysis and consequently lead to instrument-specific systematic deviations. The same effects may also result from sample and/or filter specific properties such as the presence of metals, metal oxides, oxygen donors or other substances that can catalytically affect pyrolytical processes. An example can be seen for the laser transmittance signal values of filter G, where all participants showed signs of pre-oxidation, which was not the case for the rest of the filters analysed (Figs. 11, 12). The current study gave no indication that the pre-oxidation effect is thermal protocol specific. While such comparison exercises can point out instrument specific pre-oxidation, individual users can also identify it through examination of the laser signal during sucrose solutions analysis. In this case, the laser signal will systematically ramp up during the He phase, caused by the early desorption of POC from the filter.

When the laser transmittance signal was compared among the participants a wide range in the intensity for a blank filter was noted, varying from ~ 1000 to $\sim 20\,000$. However, no systematic differences were observed in the determination of the split point after comparing weak signals with stronger ones. High or low frequency noise in the laser signal was identified for a few participants, which can cause a shift in the split point definition.

The NIOSH870 heating profile ramps up to 870 °C in 360 s during the helium phase, which appears short and aggressive when compared to EUSAAR2, which ramps up to 650 °C in 640 s (Table 2). The POC (TOT) levels were expected to be higher for NIOSH870 after the temperature offset corrections, as observed by Phuah et al. (2009) and Pavlovic et al. (2014). Nevertheless, when POC was compared between the two thermal protocols, NIOSH870 reported generally lower concentrations than EUSAAR2, even for sucrose solutions (Figs. 6–8). A possible explanation could be that lower temperature steps in the inert phase result in more OC exposed during the highest temperature step of the inert phase, and thus more POC formation (Phuah et al., 2009; Pavlovic et al., 2014). Furthermore, the longer residence time of EUSAAR2 on lower temperatures during the inert phase may increase the formation of additional POC compared to NIOSH870.

When the TOT laser signal between the two thermal protocols was compared, no clear difference on the lowest (darkest) point was observed, which is related to POC. This may suggest that not all of the OC was evolved or transformed to POC during the inert phase of EUSAAR2 and some was erroneously measured as POC and EC in the oxygen phase. For the most part, the observations associated with heating profile differences between the two thermal protocols are limited by the number of filters included in the comparison exercise, and further insight would require additional analysis on a wider selection of filters.

EC (TOT) levels were found to be lower for NIOSH870 than for EUSAAR2, explained by the differences in the heating profiles. Considering the fact that almost all analysers in this exercise had to correct their higher temperatures after calibration by significant amounts, it would be expected that different EC values would be reported after the temperature offset corrections than before, for the same analyser and sample. The magnitude of this effect would be instrument-specific and can only be evaluated by each user independently.

solutions as well as PM loaded punches of filters functioned as an efficient indicator of erroneous TC calibration and/or laboratory poor performance. Overall, outliers and deviating results were not related to the thermal protocol used.

Reported TC differed on average by only 2 % between EUSAAR2 and NIOSH870 after the implementation of the temperature calibration. EC concentrations differed by a similar ratio to ones observed previously, with $EC_{NIOSH870,TOT} = 0.80 \times EC_{EUSAAR2,TOT}$ ($R^2 = 0.96$). The TC repeatability and reproducibility (expressed as relative standard deviations) were found to be higher for EUSAAR2 (11 and 15 %) than for NIOSH870 (9.2 and 12 %). Repeatability and reproducibility for EC was found to be lower for EUSAAR2 (15 and 20 %) than for NIOSH870 (20 and 26 %). The formation of POC was found to be higher for EUSAAR2 than for NIOSH870.

Pre-oxidation was identified by low or no POC, or by an increase of the laser signal during the inert phase. Pre-oxidation was either instrument specific, originating from oxygen present in the system during the inert phase, or sample specific, due to metal oxides and/or oxygen donors present on the sample. Thermograms and laser signal plots obtained within the comparison exercise helped to clearly categorize any pre-oxidation event.

Overall, the current comparison exercise findings are two-fold. Firstly, comparison exercises that focus on laboratory performance should be implemented in laboratories' QA/QC procedures in order to reduce the likelihood of systematic errors and/or inaccuracies during ECOC analysis. And secondly, additional operational parameters and protocols should be considered for standardization or reporting, in the same way as temperature offset corrections and standardized thermal protocols have been used in the current study. A list of such parameters would include the initial laser value, POC concentration, calibration area stability, FID and laser signal plots. Actions of that kind can improve the consistency of reported EC and TC results, as well as comparability to surrogates of EC, such as black carbon and black smoke.

5 Conclusions

An ECOC comparison exercise was organized by GGD Amsterdam involving 17 laboratories. Unlike earlier comparison exercises, the participants had to perform a temperature calibration and adjust for offsets prior to analysis, for both EUSAAR2 and NIOSH870 thermal protocols. The offsets ranged from −90 to +100 °C, varying for each temperature step and instrument, pointing out the necessity of the temperature calibration. The analysis of known volumes of sucrose

Acknowledgements. The authors would like to thank all laboratories who participated in the current comparison exercise for their kind collaboration. They are particularly grateful to the providers of the HVS samples, Christoph Hüglin (EMPA, Switzerland), Kostas Eleftheriadis (Demokritos, Greece), and the department the Air Quality Monitoring Network of Amsterdam. Finally, they would like to thank Sunset Laboratory for providing the temperature calibration kits and supporting their application.

Edited by: P. Herckes

References

Adar, S. D. and Kaufman, J. D.: Cardiovascular disease and air pollutants: Evaluating and improving epidemiological data implicating traffic exposure, Inhal. Toxicol., 19, 135–149, 2007.

Baumgardner, D., Popovicheva, O., Allan, J., Bernardoni, V., Cao, J., Cavalli, F., Cozic, J., Diapouli, E., Eleftheriadis, K., Genberg, P. J., Gonzalez, C., Gysel, M., John, A., Kirchstetter, T. W., Kuhlbusch, T. A. J., Laborde, M., Lack, D., Müller, T., Niessner, R., Petzold, A., Piazzalunga, A., Putaud, J. P., Schwarz, J., Sheridan, P., Subramanian, R., Swietlicki, E., Valli, G., Vecchi, R., and Viana, M.: Soot reference materials for instrument calibration and intercomparisons: a workshop summary with recommendations, Atmos. Meas. Tech., 5, 1869–1887, doi:10.5194/amt-5-1869-2012, 2012.

Birch, M. E. and Cary, R. A.: Elemental carbon-based method for monitoring occupational exposures to particulate diesel exhaust, Aerosol Sci. Technol., 25, 221–241, 1996.

Cavalli, F., Viana, M., Yttri, K. E., Genberg, J., and Putaud, J.-P.: Toward a standardised thermal-optical protocol for measuring atmospheric organic and elemental carbon: the EUSAAR protocol, Atmos. Meas. Tech., 3, 79–89, doi:10.5194/amt-3-79-2010, 2010.

Cavalli, F., Douglas, K., and Borowiak, A.: Results of the 2nd comparison exercise for EU National Quality Reference Laboratories (AQUILA) for TC, OC and EC measurements, Ispra: Joint Research Center, 2012.

CEN TR 16243: Guide for the measurement of Elemental Carbon (EC) and Organic Carbon (OC) deposit on filters, Brussels: European Committee for Standardization, 2011.

Cheng, Y., Duan, F. K., He, K. B., Zheng, M., Du, Z. Y., Ma, Y. L., and Tan, J. H.: Intercomparison of thermal–optical methods for the determination of organic and elemental carbon: influences of aerosol composition and implications, Environ. Sci. Technol., 45, 10117–10123, 2011.

Cheng, Y., Duan, F. K., He, K. B., Du, Z. Y., Zheng, M., and Ma, Y., L.: Intercomparison of thermal-optical method with different temperature protocols: Implications from source samples and solvent extraction, Atmos. Environ., 61, 453–462, 2012.

Chow, J. C., Watson, J. G., Chen, L. W. A., Arnott, W. P., Moosmüller, H., and Fung, K.: Equivalence of elemental carbon by thermal/optical reflectance and transmittance with different temperature protocols, Environ. Sci. Technol., 38, 4414–4422, 2004.

Chow, J. C., Watson, J. G., Chen, L.-W. A., Paredes-Miranda, G., Chang, M.-C. O., Trimble, D., Fung, K. K., Zhang, H., and Zhen Yu, J.: Refining temperature measures in thermal/optical carbon analysis, Atmos. Chem. Phys., 5, 2961–2972, doi:10.5194/acp-5-2961-2005, 2005.

Chow, J. C., Watson, J. G., Chen, L. W. A., Chang, M. C. O., Robinson, N. F., Trimble, D. L., and Kohl, S. D.: The Improve-A temperature protocol for thermal/optical carbon analysis: maintaining consistency with a long-term database, J. Air Waste Manage. Assoc., 57, 1014–1023, 2007.

Chow, J. C., Watson, J. G., Robles, J., Wang, X. L., Cheng, L. W. A., Trimble, D. L., Kohl, S. D., Tropp, R. J., and Fung, K. K.: Quality assurance and quality control for thermal/optical analysis of aerosol samples for organic and elemental carbon, Anal. Bioanal. Chem., 401, 3141–3152, 2011.

Emblico, L., Cavalli, F., Hafkenscheid, T., and Borowiak, A.: Results of the first E/OC comparison exercise for EU National

Air Quality Reference Laboratories (AQUILA), Ispra: Joint Research Center, 2012.

Highwood, E. J. and Kinnersley, R. P.: When smoke gets in our eyes: The multiple impacts of atmospheric black carbon on climate, air quality and health, Environ. Int., 32, 560–566, 2006.

Hitzenberger, R., Petzold, A., Bauer, H., Ctyroky, P., Pouresmaeil, P., Laskus, L., and Puxbaum, H.: Intercomparison of thermal and optical measurement methods for elemental carbon and black carbon at an urban location, Environ. Sci. Technol., 40, 6377–6383, 2006.

IPCC: Climate Change 2007: Impacts, Adaptation and Vulnerability. Contribution of Working Group II to the Fourth Assessment Report of the Intergovernmental Panel on Climate Change, edited by: Parry, M. L., Canziani, O. F., Palutikof, J. P., van der Linden, P. J., and Hanson, C. E., Cambridge University Press, Cambridge, UK, 976 pp., 2007.

ISO 13528: Statistical methods for use in proficiency testing by inter-laboratory comparisons, ISO, Geneva, 2005.

ISO 5725-2: Accuracy (trueness and precision) of measurement methods and results – Part 2: Basic method for the determination of repeatability and reproducibility of a standard measurement method, ISO, Geneva, 1994.

Jacobson, M. Z.: Strong radiative heating due to the mixing state of black carbon in atmospheric aerosols, Nature, 409, 695–697, 2001.

Janssen, N., Hoek, G., Simic-Lawson, M., Fischer, P., van Bree, L., ten Brink, H., Keuken, M., Atkinson, R., Anderson, H. R., Brunekreef, B., and Cassee, F. R.: Black carbon as an additional indicator of the adverse health effects of airborne particles compared with PM_{10} and $PM_{2.5}$, Environ. Health Perspect., 119, 1691–1699, 2011.

Janssen, N., Gerlofs-Nijland, M., Lanki, T., Salonen, R., Cassee, F., Hoek, G., Fischer, P., Brunekreef, B., and Krzyzanowski, M.: Health effects of black carbon, Copenhagen, Denmark: WHO, Regional Office for Europe, 2012.

Lena, S. T., Carter, M., Holguin-Veras, J., and Kinney, P. L.: Elemental carbon and PM2.5 levels in an urban community heavily impacted by truck traffic, Environ. Health Perspect., 110, 1009–1015, 2002.

Maenhaut, W., Claeys, M., Vercauteren, J., and Roekens, E.: Comparison of reflectance and transmission in EC/OC measurements of filter samples from Flanders, Belgium. Abstract Book of the 10th International Conference on Carbonaceous Particles in the Atmosphere (ICCPA), 26–29 June 2011, Vienna, Austria, Abstract F-6, 2011.

Panteliadis, P.: EC/OC Interlaboratory comparison measurements. 2009 Report, GGD/LO 09-117, Public Health Service Amsterdam, Department of Air Quality, 2009a.

Panteliadis, P.: EC-OC Measurements: Comparison of reflectance and transmittance techniques, GGD/LO 09-1135, Public Health Service Amsterdam, Department of Air Quality, 2009b.

Panteliadis, P.: EC/OC Interlaboratory comparison. 2010 Report, GGD/LO11-1112, Public Health Service Amsterdam, Department of Air Quality, 2011.

Panteliadis, P., Strak, M., Hoek, G., Weijers, E., van der Zee, S., and Dijkema, M.: Implementation of a low emission zone and evaluation of effects on air quality by long-term monitoring, Atmos. Environ., 86, 113–119, 2014.

Pavlovic, J., Kinsey, J. S., and Hays, M. D.: The influence of temperature calibration on the OC-EC results from a dual optics thermal carbon analyzer, Atmos. Meas. Tech. Discuss., 7, 3321–3348, doi:10.5194/amtd-7-3321-2014, 2014.

Phuah, C. H., Peterson, M. R., Richards, M. R., Turner, J. H., and Dillner, A. M.: A temperature calibration procedure for the Sunset Laboratory carbon aerosol analysis lab instrument, Aerosol Sci. Technol., 43, 1013–1021, 2009.

Piazzalunga, A., Bernardoni, V., Fermo, P., Valli, G., and Vecchi, R.: Technical Note: On the effect of water-soluble compounds removal on EC quantification by TOT analysis in urban aerosol samples, Atmos. Chem. Phys., 11, 10193–10203, doi:10.5194/acp-11-10193-2011, 2011.

Qadir, R. M., Abbaszade, G., Schnelle-Kreis, J., Chow, J. C., and Zimmermann, R.: Concentrations and source contributions of particulate organic matter before and after implementation of a low emission zone in Munich, Germany, Environ. Pollut., 175, 158–167, 2013.

Ramanathan, V. and Carmichael, G.: Global and regional climate changes due to black carbon, Nature Geosci., 1, 221–227, 2008.

Schauer, J.: Evaluation of elemental carbon as a marker for diesel particulate matter, J. Expo. Sci. Environ. Epidemiol., 13, 443–453, 2003.

Schauer, J., Mader, B., Deminter, J., Heidemann, G., Bae, M., Seinfeld, H., Flagan, R. C., Cary, R. A., Smith, D., Huebert, B. J., Bertram, T., Howell, S., Kline, J. T., Quinn, P., Bates, T., Turpin, B., Lim, H. J., Yu, J. Z., Yang, H., and Keywood, M.: ACE-Asia intercomparison of a thermal-optical method for the determination of particle-phase organic and elemental carbon, Eviron. Sci. Technol., 37, 993–1001, 2003.

Schmid, H., Laskus, L., Abraham, H. J., Baltensperger, U., Lavanchy, V., Bizjak, M., Burba, P., Cachier, H., Crow, D., Chow, J., Gnauk, T., Even, A., ten Brink, H. M., Giesen, K. P., Hitzenberger, R., Hueglin, C., Maenhaut, W., Pio, C., Carvalho, A., Putaud, J. P., Toom-Sauntry, D., and Puxbaum, H.: Results of the "carbon conference" international aerosol carbon round robin test stage I, Atmos. Environ., 35, 2111–2121, 2001.

Sciare, J., Cachier, H., Oikonomou, K., Ausset, P., Sarda-Estève, R., and Mihalopoulos, N.: Characterization of carbonaceous aerosols during the MINOS campaign in Crete, July–August 2001: a multi-analytical approach, Atmos. Chem. Phys., 3, 1743–1757, doi:10.5194/acp-3-1743-2003, 2003.

ten Brink, H., Maenhaut, W., Hitzenberger, R., Gnauk, T., Spindler, G., Even, A., Chi, X., Bauer, H., Puxbaum, H., Putaud J.-P., Tursic, J., and Berner, A.: INTERCOMP2000: the comparability of methods in use in Europe for measuring the carbon content of aerosol, Atmos. Environ., 38, 6507–6519, 2004.

Wang, Y., Chung, A., and Paulson, S. E.: The effect of metal salts on quantification of elemental and organic carbon in diesel exhaust particles using thermal-optical evolved gas analysis, Atmos. Chem. Phys., 10, 11447–11457, doi:10.5194/acp-10-11447-2010, 2010.

Watson, J. G., Chow, J. C., and Chen, L. W. A.: Summary of organic and elemental carbon/black carbon analysis methods and intercomparisons, Aerosol Air Qual. Res., 5, 65–102, 2005.

Zhi, G. R., Chen, Y. J., Sun, J. Y., Chen, L. G., Tian, W. J., Duan, J. C., Zhang, G., Chai, F. H., Sheng, G. Y., and Fu, J. M.: Harmonizing aerosol carbon measurements between two conventional thermal/optical analysis methods, Environ. Sci. Technol., 7, 2902–2908, 2011.

Permissions

The contributors of this book come from diverse backgrounds, making this book a truly international effort. This book will bring forth new frontiers with its revolutionizing research information and detailed analysis of the nascent developments around the world.

We would like to thank all the contributing authors for lending their expertise to make the book truly unique. They have played a crucial role in the development of this book. Without their invaluable contributions this book wouldn't have been possible. They have made vital efforts to compile up to date information on the varied aspects of this subject to make this book a valuable addition to the collection of many professionals and students.

This book was conceptualized with the vision of imparting up-to-date information and advanced data in this field. To ensure the same, a matchless editorial board was set up. Every individual on the board went through rigorous rounds of assessment to prove their worth. After which they invested a large part of their time researching and compiling the most relevant data for our readers.

The editorial board has been involved in producing this book since its inception. They have spent rigorous hours researching and exploring the diverse topics which have resulted in the successful publishing of this book. They have passed on their knowledge of decades through this book. To expedite this challenging task, the publisher supported the team at every step. A small team of assistant editors was also appointed to further simplify the editing procedure and attain best results for the readers.

Apart from the editorial board, the designing team has also invested a significant amount of their time in understanding the subject and creating the most relevant covers. They scrutinized every image to scout for the most suitable representation of the subject and create an appropriate cover for the book.

The publishing team has been an ardent support to the editorial, designing and production team. Their endless efforts to recruit the best for this project, has resulted in the accomplishment of this book. They are a veteran in the field of academics and their pool of knowledge is as vast as their experience in printing. Their expertise and guidance has proved useful at every step. Their uncompromising quality standards have made this book an exceptional effort. Their encouragement from time to time has been an inspiration for everyone.

The publisher and the editorial board hope that this book will prove to be a valuable piece of knowledge for researchers, students, practitioners and scholars across the globe.

List of Contributors

S. A. P. de Jong
Faculty of Civil Engineering and Geosciences, Delft University of Technology, Delft, the Netherlands

J. D. Slingerland
Faculty of Civil Engineering and Geosciences, Delft University of Technology, Delft, the Netherlands

N. C. van de Giesen
Faculty of Civil Engineering and Geosciences, Delft University of Technology, Delft, the Netherlands

K. Ramesh
Department of Computer Applications, Anna University, Regional Center, Tirunelveli, Tamil Nadu 627 005, India

A. P. Kesarkar
National Atmospheric Research Laboratory, Gadanki 517 112, Chittoor District, Andhra Pradesh, India

J. Bhate
National Atmospheric Research Laboratory, Gadanki 517 112, Chittoor District, Andhra Pradesh, India

M. Venkat Ratnam
National Atmospheric Research Laboratory, Gadanki 517 112, Chittoor District, Andhra Pradesh, India

A. Jayaraman
National Atmospheric Research Laboratory, Gadanki 517 112, Chittoor District, Andhra Pradesh, India

L.-W. A. Chen
Department of Environmental and Occupational Health, University of Nevada, Las Vegas, Nevada 89154, USA
Division of Atmospheric Sciences, Desert Research Institute, Reno, Nevada 89512, USA
Key Laboratory of Aerosol Science & Technology, SKLLQG, Institute of Earth Environment, Chinese Academy of Sciences, Xi'an, China

J. C. Chow
Division of Atmospheric Sciences, Desert Research Institute, Reno, Nevada 89512, USA
Key Laboratory of Aerosol Science & Technology, SKLLQG, Institute of Earth Environment, Chinese Academy of Sciences, Xi'an, China

X. L. Wang
Key Laboratory of Aerosol Science & Technology, SKLLQG, Institute of Earth Environment, Chinese Academy of Sciences, Xi'an, China

J. A. Robles
Key Laboratory of Aerosol Science & Technology, SKLLQG, Institute of Earth Environment, Chinese Academy of Sciences, Xi'an, China

B. J. Sumlin
Key Laboratory of Aerosol Science & Technology, SKLLQG, Institute of Earth Environment, Chinese Academy of Sciences, Xi'an, China

D. H. Lowenthal
Key Laboratory of Aerosol Science & Technology, SKLLQG, Institute of Earth Environment, Chinese Academy of Sciences, Xi'an, China

R. Zimmermann
Joint Mass Spectrometry Centre, Chair of Analytical Chemistry, Institute of Chemistry, University of Rostock, Rostock, Germany

J. G. Watson
Division of Atmospheric Sciences, Desert Research Institute, Reno, Nevada 89512, USA
Key Laboratory of Aerosol Science & Technology, SKLLQG, Institute of Earth Environment, Chinese Academy of Sciences, Xi'an, China

D. Butterfield
National Physical Laboratory, Hampton Road, Teddington, Middlesex, TW11 0LW, UK

T. Gardiner
National Physical Laboratory, Hampton Road, Teddington, Middlesex, TW11 0LW, UK

T. Fauchez
Laboratoire d'Optique Atmosphérique, Université Lille 1, Villeneuve d'Ascq, France

P. Dubuisson
Laboratoire d'Optique Atmosphérique, Université Lille 1, Villeneuve d'Ascq, France

C. Cornet
Laboratoire d'Optique Atmosphérique, Université Lille 1, Villeneuve d'Ascq, France

F. Szczap
Laboratoire de Météorologie Physique, Université Blaise Pascal, Clermont Ferrand, France

A. Garnier
Science Systems and Applications, Inc., Hampton, Virginia, USA
NASA Langley Research Center, Hampton, Virginia, USA

J. Pelon
Laboratoire Atmosphères, Milieux, Observations Spatiales, UPMC-UVSQ-CNRS, Paris, France

K. Meyer
Goddard Earth Sciences Technology and Research (GESTAR), Universities Space Research Association, Columbia, Maryland, USA
NASA Goddard Space Flight Center, Greenbelt, Maryland, USA

E. Defer
LERMA, UMR8112, Observatoire de Paris & CNRS, Paris, France

J.-P. Pinty
LA, UMR5560, Université de Toulouse & CNRS, Toulouse, France

S. Coquillat
LA, UMR5560, Université de Toulouse & CNRS, Toulouse, France

J.-M. Martin
LA, UMR5560, Université de Toulouse & CNRS, Toulouse, France

S. Prieur
LA, UMR5560, Université de Toulouse & CNRS, Toulouse, France

S. Soula
LA, UMR5560, Université de Toulouse & CNRS, Toulouse, France

E. Richard
LA, UMR5560, Université de Toulouse & CNRS, Toulouse, France

W. Rison
NMT, Socorro, New Mexico, USA

P. Krehbiel
NMT, Socorro, New Mexico, USA

R. Thomas
NMT, Socorro, New Mexico, USA

D. Rodeheffer
NMT, Socorro, New Mexico, USA

C. Vergeiner
Institute of High Voltage Engineering and System Performance, TU Graz, Graz, Austria

F. Malaterre,
Météorage, Pau, France

S. Pedeboy
Météorage, Pau, France

W. Schulz
OVE-ALDIS, Vienna, Austria

T. Farges
CEA, DAM, DIF, Arpajon, France

L.-J. Gallin
CEA, DAM, DIF, Arpajon, France

P. Ortéga
GePaSUD, UPF, Faa'a, Tahiti, French Polynesia

J.-F. Ribaud
CNRM-GAME, UMR3589, Météo-France & CNRS, Toulouse, France

G. Anderson
UK Met Office, Exeter, UK

H.-D. Betz
nowcast, Garching, Germany

B. Meneux
nowcast, Garching, Germany

V. Kotroni
NOA, Athens, Greece

K. Lagouvardos
NOA, Athens, Greece

S. Roos
Météo France, Nîmes, France

V. Ducrocq
CNRM-GAME, UMR3589, Météo-France & CNRS, Toulouse, France

O. Roussot
CNRM-GAME, UMR3589, Météo-France & CNRS, Toulouse, France

L. Labatut
CNRM-GAME, UMR3589, Météo-France & CNRS, Toulouse, France

G. Molinié
LTHE, Grenoble, France

C. Budke
Faculty of Chemistry, Bielefeld University, Universitätsstraße 25, 33615 Bielefeld, Germany

T. Koop
Faculty of Chemistry, Bielefeld University, Universitätsstraße 25, 33615 Bielefeld, Germany

P. Panteliadis
Municipal Health Service (GGD) Amsterdam, Department of Air Quality, Amsterdam, the Netherlands

T. Hafkenscheid
National Institute for Public Health and the Environment, Bilthoven, the Netherlands

B. Cary
Sunset Laboratory Inc, Tigard, Oregon, USA

E. Diapouli
National Center for Scientific Research "Demokritos", Institute of Nuclear & Radiological Sciences & Technology, Energy & Safety, Athens, Greece

A. Fischer
EMPA – Swiss Federal Laboratories for Materials Science and Technology, Duebendorf, Switzerland

O. Favez
INERIS, Verneuil-en-Halatte, France

P. Quincey
National Physical Laboratory, Teddington, UK

M. Viana
Institute for Environmental Assessment and Water Research (IDAEA-CSIC), Barcelona, Spain

R. Hitzenberger
Aerosolphysics and Environmental Physics, Faculty of Physics, Vienna, Austria

R. Vecchi
Department of Physics, Università degli Studi di Milano, Milan, Italy

D. Saraga
Demokritos, National Center for Scientific Research, Environmental Research Laboratory, Athens, Greece

J. Sciare
Laboratoire des Sciences du Climat et de l'Environnement (LSCE), CEA-CNRS-UVSQ, Gif-sur-Yvette, France

J. L. Jaffrezo
Univ. Grenoble Alpes, CNRS, LGGE, 38000 Grenoble, France

A. John
Institute for Energy and Environmental Technology e.V. Air Quality & Sustainable Nanotechnology Division, Duisburg, Germany

J. Schwarz
Institute of Chemical Process Fundamentals AS CR, Prague, Czech Republic

M. Giannoni
Istituto Nazionale di Fisica Nucleare (INFN), Sezione di Firenze, Florence, Italy

J. Novak
Czech Hydrometeorological Institute, Prague, Czech Republic

A. Karanasiou
Institute for Environmental Assessment and Water Research (IDAEA-CSIC), Barcelona, Spain

P. Fermo
Department of Chemistry, Università degli Studi di Milano, Milan, Italy

W. Maenhaut
Department of Analytical Chemistry, Ghent University, Gent, 9000, Belgium

Printed in the USA
CPSIA information can be obtained
at www.ICGtesting.com
JSHW051447221024
72173JS00006B/1606

9 781682 860007